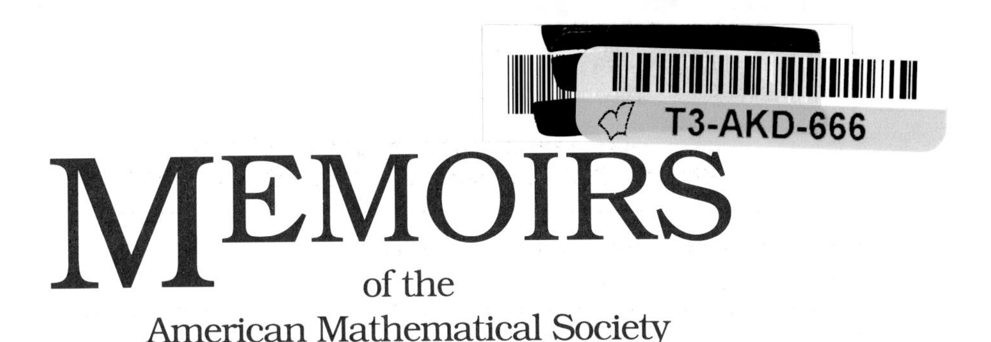

MEMOIRS
of the
American Mathematical Society

Number 462

Singular Unitary Representations and Discrete Series for Indefinite Stiefel Manifolds $U(p,q;\mathbb{F})/U(p-m,q;\mathbb{F})$

Toshiyuki Kobayashi

January 1992 • Volume 95 • Number 462 (end of volume) • ISSN 0065-9266

American Mathematical Society
Providence, Rhode Island

1991 *Mathematics Subject Classification.*
Primary 22E46; Secondary 43A85, 22E45, 22E47.

Library of Congress Cataloging-in-Publication Data

Kobayashi, Toshiyuki, 1962–
Singular unitary representations and discrete series for indefinite Stiefel manifolds $U(p, q; \mathbb{F})/U(p - m, q; \mathbb{F})$/Toshiyuki Kobayashi.
p. cm. – (Memoirs of the American Mathematical Society, ISSN 0065-9266; no. 462)
Includes bibliographical references.
ISBN 0-8218-2524-0
1. Semisimple Lie groups. 2. Representations of groups. 3. Harmonic analysis. 4. Stiefel manifolds. I. Title. II. Series.
QA3.A57 no. 462
[QA387]
510 s–dc20 91-36299
[512′.55] CIP

Subscriptions and orders for publications of the American Mathematical Society should be addressed to American Mathematical Society, Box 1571, Annex Station, Providence, RI 02901-1571. *All orders must be accompanied by payment.* Other correspondence should be addressed to Box 6248, Providence, RI 02940-6248.

SUBSCRIPTION INFORMATION. The 1992 subscription begins with Number 459 and consists of six mailings, each containing one or more numbers. Subscription prices for 1992 are $292 list, $234 institutional member. A late charge of 10% of the subscription price will be imposed on orders received from nonmembers after January 1 of the subscription year. Subscribers outside the United States and India must pay a postage surcharge of $25; subscribers in India must pay a postage surcharge of $43. Expedited delivery to destinations in North America $30; elsewhere $82. Each number may be ordered separately; *please specify number* when ordering an individual number. For prices and titles of recently released numbers, see the New Publications sections of the NOTICES of the American Mathematical Society.

BACK NUMBER INFORMATION. For back issues see the AMS Catalogue of Publications.

MEMOIRS of the American Mathematical Society (ISSN 0065-9266) is published bimonthly (each volume consisting usually of more than one number) by the American Mathematical Society at 201 Charles Street, Providence, Rhode Island 02904-2213. Second Class postage paid at Providence, Rhode Island 02940-6248. Postmaster: Send address changes to Memoirs of the American Mathematical Society, American Mathematical Society, Box 6248, Providence, RI 02940-6248.

Printed in the United States of America.
The paper used in this book is acid-free and falls within the guidelines established to ensure permanence and durability. ♾

10 9 8 7 6 5 4 3 2 1 97 96 95 94 93 92

Contents

Abstract

This paper treats the relatively singular part of the unitary dual of pseudo-orthogonal groups over $\mathbb{F} = \mathbb{R}$, $\mathbb{C}$ and $\mathbb{H}$. These representations arise from discrete series for indefinite Stiefel manifolds $U(p,q;\mathbb{F})/U(p-m,q;\mathbb{F})$ $(2m \leq p)$. Thanks to the duality theorem between $\mathcal{D}$-module construction and Zuckerman's derived functor modules (ZDF-modules), these discrete series are naturally described in terms of ZDF-modules with possibly singular parameters. Some techniques including a new K-type formula are offered to find the explicit condition deciding whether the corresponding ZDF-modules $\mathcal{R}_{\mathfrak{q}}^{S}(\mathbb{C}_\lambda)$ vanish or not. We also investigate the irreducibility and pairwise inequivalence among these ZDF-modules. Although our concern is limited to the discrete series, our approach is purely algebraic and applicable to a less special setting. It is an interesting phenomenon that our discrete series sometimes give a sharper condition for unitarizability of ZDF-modules than those given by Vogan (1984) algebraically. This phenomenon does not occur in the case of discrete series for group manifolds or semisimple symmetric spaces.

keywords and phrases: unitary representations of semisimple Lie groups, Zuckerman's derived functor modules, discrete series, symmetric spaces, pseudo-orthogonal groups

§0. Introduction

Interesting classes of $(\mathfrak{g}, K)$-modules are often described naturally in terms of cohomologically induced representations in various settings such as unitary highest weight modules, the theory of dual reductive pairs, discrete series for semisimple symmetric spaces, and etc. These have been stimulating the study of algebraic properties of derived functor modules. Now an almost satisfactory theory on derived functor modules including a functorial property about unitarizability, has been developed in the 'good' range of parameters. However, actual interesting families of unitary representations contain possibly singular parameters, so we are still faced with subtle problems such as finding conditions for non-vanishing, irreducibility, or pairwise inequivalence of these $(\mathfrak{g}, K)$-modules (see Problem(0.6)$'$). This paper treats such delicate algebraic properties of Zuckerman's derived functor modules with singular parameters which arise from discrete series for indefinite Stiefel manifolds $U(p,q;\mathbb{F})/U(p-m,q;\mathbb{F})$ $(p \geq 2m,\ \mathbb{F} = \mathbb{R}, \mathbb{C}$ or $\mathbb{H})$. Some of them are isomorphic to "unipotent" representations in the philosophy of Arthur (e.g. [1], §5), and some are out of the range given by Vogan [29] for his unitarizability theorem. We should also remark that the unitary dual of a pseudo-orthogonal group $U(p,q;\mathbb{F})$ has not been classified yet, except in lower rank cases.

Let G be a connected real reductive linear Lie group and H be a closed subgroup which is reductive in G. A homogeneous space G/H carries a G-invariant measure, so we have a natural unitary representation of G on the Hilbert space $L^2(G/H)$. Harish-Chandra modules of finite length realized in $L^2(G/H)$ are called *discrete series* for G/H. These play a fundamental role in harmonic analysis on G/H. If H is noncompact,

Received by the editors 6 March, 1990 and, in revised form 30 April 1990.
Partially supported by Grant-in-Aid for Scientific Research (No.01740081).

discrete series for G/H do not necessarily enter the Plancherel decomposition in $L^2(G)$ and then should correspond to a relatively singular part of the unitary dual of G.

If H is an open subgroup of the fix point group of an involutive automorphism σ of G, G/H is a typical example of a homogeneous space of reductive type and called a *semisimple symmetric space.* Much progress on discrete series for semisimple symmetric spaces has been made in the last ten years (see e.g. [2], [8], [9], [20], [21], [24], [33]). In particular, discrete series for G/H is not empty iff

$$\operatorname{rank} G/H = \operatorname{rank} K/H \cap K, \tag{0.1}$$

where K is a σ-stable maximal compact subgroup in G.

Suppose that H has a direct decomposition $H_1 \times H_2$ with H_1 compact. Correspondingly to the G-equivariant H_1-principal bundle

$$H_1 \to G/H_2 \to G/H,$$

we have

$$L^2(G/H) = L^2(G/H; \mathbf{1}) \hookrightarrow L^2(G/H_2) \simeq \bigoplus_{\tau} \dim \tau \, L^2(G/H; \tau),$$

where the sum is taken over the unitary dual of the compact group H_1 and $L^2(G/H;\tau)$ denotes the space of the square integrable sections of the vector bundle over G/H associated to τ.

Throughout this section, we will write G/H, G/H_2 as a symbol of semisimple symmetric spaces with the rank condition (0.1), homogeneous spaces in the above setting, respectively. Since we allow the case $H_1 = \{e\}$ and $H_2 = H$, G/H_2 represents a wider class of homogeneous spaces than G/H in this notation. Typical examples of $G \supset H = H_1 \times H_2$ are given in (0.10), and if G/H is any other classical irreducible semisimple symmetric space then H_1 is one dimensional (or zero dimensional).

It is natural to extend the study of harmonic analysis from $L^2(G/H)$ to $L^2(G/H_2)$, in which more general unitary representations of G are expected to be realized. Discrete series for G/H_2 have not been studied very well before, except in Schlichtkrull's original work [23] generalizing Flensted-Jensen's construction [8]. Building on work of Schlichtkrull and Oshima-Matsuki [20], we construct a wider class of discrete series for G/H_2 whose parameters are possibly more singular.

Roughly, our construction is as follows: In a complexification $G_{\mathbb{C}}$ of G, other real forms $K^r \subset G^r \supset H^r \approx {H_1}^r \times {H_2}^r$ of $K_{\mathbb{C}} \subset G_{\mathbb{C}} \supset H_{\mathbb{C}} \approx H_{1\mathbb{C}} \times H_{2\mathbb{C}}$ are defined so that H^r is compact [9]. Let $P^r = M^rA^rN^r$ be a minimal parabolic subgroup of G^r, $\rho \equiv \rho(\mathfrak{n}^r)$ be half the sum of roots as usual, and X_j ($1 \le j \le l$, finite) be the closed K^r-orbits in G^r/P^r. Denote by $\mathcal{A}$ the sheaf of analytic functions, by $\mathcal{B}$ that of hyperfunctions. For $(\delta, V) \in \widehat{M^r}$, $\nu \in \widehat{A^r}$, we define $\mathcal{F}$ to be the $\mathcal{A}$ or $\mathcal{B}$ valued principal series by

$$\mathcal{F}(G^r/P^r; \delta \otimes \nu) := \{F \in \mathcal{F}(G^r; V)\,;\ F(gman) = \delta(m)^{-1}a^{-\nu-\rho}F(g) \quad \text{for } m \in M^r,\ a \in A^r,\ n \in N^r\}.$$

Then naturally we have a G^r-invariant nondegenerate bilinear form

$$B\colon \mathcal{B}(G^r/P^r; \delta^* \otimes (-\nu)) \times \mathcal{A}(G^r/P^r; \delta \otimes \nu) \to \mathbb{C}, \tag{0.2}$$

by $B(F, v) := \int_{H^r} \langle F(k), v(k)\rangle\, dk$. If v is a cyclic vector in $\mathcal{A}(G^r/P^r; \delta \otimes \nu)$ with trivial ${H_2}^r$-action, then B induces an injective mapping

$$\mathcal{P}_v\colon \mathcal{B}(G^r/P^r; \delta^* \otimes (-\nu)) \to \mathcal{A}(G^r/{H_2}^r), \tag{0.3}$$

by $\mathcal{P}_v(F)(g) := B(F(g\cdot\ ), v) = B(F, g\cdot v)$. Then $\mathcal{P}_v$ is an analogue of the *Poisson transform* which respects the left G^r-action. Let

$$\mathcal{B}^j_{K^r}(\delta \otimes \nu) := \{F \in \mathcal{B}(G^r/P^r; \delta \otimes \nu)\,;\ \operatorname{supp} F \subset X_j,\ F \text{ transforms according to finite dimensional representations of } K_{\mathbb{C}} \text{ under the action of } K^r\}.$$

Now we are ready to state a construction of discrete series for G/H_2:

Proposition 0.4. *Assume that*

(0.5)(a) $\mathcal{B}^j_{K^r}(G^r/P^r; \delta^* \otimes (-\nu)) \neq 0$,

(0.5)(b) $\mathcal{A}(G^r/P^r; \delta \otimes \nu)$ *contains a* ${H_2}^r$*-fixed cyclic vector* v,

(0.5)(c) $\langle \nu, \alpha \rangle > 0$ *for any* $\alpha \in \Sigma(\mathfrak{g}^r, \mathfrak{a}^r)$.

Then the image of $\mathcal{P}_v\left(\mathcal{B}^j_{K^r}(\delta^* \otimes (-\nu))\right)$ $(1 \leq j \leq l)$ *by Flensted-Jensen duality (a kind of holomorphic continuation) gives discrete series for* $L^2(G/H_2)$.

We remark that if $H_1 = \{e\}, H_2 = H$ and if $\delta = 1$, then the above construction exhausts all discrete series for semisimple symmetric spaces G/H (in particular, those for group manifolds $G \simeq G \times G/\operatorname{diag}(G)$) (see Fact(1.5.1),(1.5.2)).

Since the proof of Proposition(0.4) is similar to [23] and [20], and since the proof requires a lot of analytic notations in estimating square integrability, we do not give it in this paper. Instead, we investigate algebraic properties of these unitary representations realized in $L^2(G/H_2)$. First of all, we must find the explicit conditions for (0.5)(a),(b), which are satisfied if the parameter ν is sufficiently 'regular'.

To obtain the explicit condition for (0.5)(b) is non-trivial but fairly easy: It is well known that this condition is automatically satisfied under (0.5)(c) if ${H_1}^r = \{e\}$ due to Helgason, Kostant [10], [16]. This condition is also known if $\dim {H_1}^r = 1$ (see [19], Theorem 1.2, see also an explicit c-function obtained in [25](with a miswriting in $Sp(n, \mathbb{R})$ case), [26]). For a general ${H_1}^r$, we can calculate it by standard arguments using Gindikin-Karpalevič method and the explicit knowledge of the socle filtration of principal series for rank one simple Lie groups (see [5]). The arguments are independent and we shall report it in another paper.

The most important and non-trivial part is (0.5)(a). So our interest in this paper will be concentrated on the following algebraic properties involving (0.5)(a):

Problem 0.6. Find each of the explicit conditions, deciding whether or not, $(\mathfrak{g}, K^\tau)$-modules $\mathcal{B}^j_{K^\tau}(\delta^* \otimes (-\nu))$ $(1 \le j \le l)$ vanish, are irreducible, are pairwise inequivalent and etc.

Thanks to the duality theorem due to Hecht, Miličic, Schmid and Wolf ([4], [11], see also [2] II.6), the modules $\mathcal{B}^j_{K^\tau}(\delta^* \otimes (-\nu))$ are isomorphic to the duals of a family of $(\mathfrak{g}, K)$-modules defined by Zuckerman. Then we may reformulate Problem (0.6) into a more familiar one:

Problem 0.6′. Find an explicit condition guaranteeing

i) $\mathcal{R}^S_{\mathfrak{q}}(\mathbb{C}_\lambda) \neq 0$ if $S := \dfrac{1}{2}\dim(K/L\cap K)$.

ii) $\mathcal{R}^j_{\mathfrak{q}}(\mathbb{C}_\lambda) = 0$ for any $j \neq S$.

iii) $\mathcal{R}^S_{\mathfrak{q}}(\mathbb{C}_\lambda)$ is irreducible or zero.

iv) $\mathcal{R}^S_{\mathfrak{q}}(\mathbb{C}_\lambda) \simeq \mathcal{R}^{S'}_{\mathfrak{q}'}(\mathbb{C}_{\lambda'})$.

Here $\mathfrak{q} = \mathfrak{l} + \mathfrak{u}$ is a θ-stable parabolic subalgebra, $\mathbb{C}_\lambda$ is a metapletic $(\mathfrak{l}, (L\cap K)^\sim)$ unitary character, and $\mathcal{R}^j_{\mathfrak{q}}(\mathbb{C}_\lambda)$ $(j \in \mathbb{N})$ are Zuckerman's cohomologically induced $(\mathfrak{g}, K)$-modules (see §1.3 for our normalization).

Remark 0.7. Our discrete series for G/H_2 are originally the dual of cohomologically induced representations from finite dimensional representations according to $\delta \in \widehat{M^\tau}$. If we use further induction by stages and the Borel-Weil-Bott theorem for compact groups, we reduce them to those induced from characters.
Let $\mathfrak{t}_0$ be a maximal abelian subalgebra of $\mathfrak{k}_0{}^{\perp(\mathfrak{h}_2)_0}$, a subspace of $\mathfrak{k}_0$ orthogonal to $(\mathfrak{h}_2)_0$. Then the discrete series constructed in Proposition(0.4) correspond to the modules $\mathcal{R}^S_{\mathfrak{q}}(\mathbb{C}_\lambda)$ where $\mathfrak{q}$ is defined by a generic element of $\mathfrak{t}_0$. The choice of closed K^τ-orbits X_j $(1 \le j \le l)$ corresponds to the choice of polarizations. We will find that the conditions (0.5)(a),(b) and (c) impose on λ some mild positivity conditions with respect to $\mathfrak{q}$ (see List(0.9), Theorems in §2).

The difficulty in Problem(0.6)′ arises from singular parameters. Let us explain it. Suppose that $\mathfrak{q} = \mathfrak{l} + \mathfrak{u}$ is a θ-stable parabolic subalgebra and let $\mathbb{C}_\lambda$ be a metapletic $(\mathfrak{l}, (L \cap K)^\sim)$-module. Three kinds of dominance conditions on λ are defined:

(0.8)(a) $\lambda + \rho_{\mathfrak{l}}$ is dominant for $\mathfrak{u}$,

(0.8)(b) λ is dominant for $\mathfrak{u}$,

(0.8)(c) $\mu_\lambda := \lambda + \rho(\mathfrak{u}) - 2\rho(\mathfrak{u} \cap \mathfrak{k})$ is $\Delta^+(\mathfrak{k})$-dominant.

We say (0.8)(a) (weekly) *good* and (0.8)(b) (weakly) *fair* according to [33] (more precisely, see Definition(1.2.1)).

General theory guarantees nice behaviors of the modules $\mathcal{R}_{\mathfrak{q}}^j(\mathbb{C}_\lambda)$ when $\mathbb{C}_\lambda$ is in the good range where $\mathcal{R}_{\mathfrak{q}}^j(\mathbb{C}_\lambda)$ has a regular $\mathfrak{Z}(\mathfrak{g})$-infinitesimal character (see Fact(1.4.1)). Indeed, suppose that a unitary character $\mathbb{C}_\lambda$ is in the good range. Then,

$$\mathcal{R}_{\mathfrak{q}}^j(\mathbb{C}_\lambda) = 0 \qquad \text{if } j \neq S,$$

$$\mathcal{R}_{\mathfrak{q}}^S(\mathbb{C}_\lambda) \text{ is nonzero, irreducible, and unitarizable.}$$

Suppose one allows $\mathbb{C}_\lambda$ to be in the (weakly) fair range, the cohomology group $\mathcal{R}_{\mathfrak{q}}^S(\mathbb{C}_\lambda)$ tends to be reducible or vanish (see Fact(1.4.2)). The condition (0.8)(c) is sufficient for the non-vanishing of $\mathcal{R}_{\mathfrak{q}}^S(\mathbb{C}_\lambda)$. The normality of the moment map (see §6) is sufficient for the irreducibility (or vanishing) of $\mathcal{R}_{\mathfrak{q}}^S(\mathbb{C}_\lambda)$. For example, these sufficient conditions apply to Proposition 6.41 in [32] which is used in classifying the unitary dual of $GL(2n, \mathbb{R})$. (This part of the book [32] suggests some importance of the study in the strip: fair but not good.) However, unfortunately none of these conditions is necessary in general.

Suppose one allows $\mathbb{C}_\lambda$ to wander outside the (weakly) fair range, the cohomology group $\mathcal{R}_{\mathfrak{q}}^S(\mathbb{C}_\lambda)$ may be non-unitary, and cohomology may turn up in other dimensions as well.

Here is a sketchy list of the ranges of the parameters of the discrete series constructed in Proposition(0.4) in terms of Zuckerman's derived functor modules (see Remark(0.7) for precise meaning). Semisimple Lie groups $G \simeq G \times G/\operatorname{diag}(G)$, semisimple symmetric spaces G/H, and our homogeneous spaces G/H_2 are assumed to satisfy the rank condition $\operatorname{rank} G = \operatorname{rank} K$, (0.1), and (0.1) respectively.

the parameter of discrete series $\mathcal{R}_{\mathfrak{q}}^S(\mathbb{C}_\lambda)$			
	G	G/H	G/H_2
$\mathbb{C}_\lambda$ is fair	Yes	Yes	No
$\mathbb{C}_\lambda$ is good	Yes	No	No
μ_λ is $\Delta^+(\mathfrak{k})$-dominant	Yes	No	No

List 0.9.

Here "Yes" means that the corresponding condition is always satisfied, and otherwise "No".

Recall that the construction in Proposition(0.4) exhausts all discrete series for G and G/H (see Fact(1.5.1),(1.5.2)). Counter examples (i.e. "No") in List(0.9) for G/H_2 are given as follows:

Suppose that G/H_2 is an indefinite Stiefel manifold

$$G/H_2 = U(p,q;\mathbb{F})/U(p-m,q;\mathbb{F}), \qquad (p \geq 2m, \quad \mathbb{F} = \mathbb{R}, \mathbb{C} \text{ or } \mathbb{H}), \tag{0.10}$$

which is one of the most interesting and typical homogeneous spaces in our setting. (In this case $H_1 = U(m;\mathbb{F})$.) Then Theorem $1 \sim 3$ in §2 says that the discrete series for G/H_2 in Proposition(0.4) contains a module $\mathcal{R}_{\mathfrak{q}}^S(\mathbb{C}_\lambda)$ such that

i) $\mathbb{C}_\lambda$ is *outside* the weakly fair range when $\mathbb{F} = \mathbb{C}$ or $\mathbb{H}$ with any p, q, m.

ii) μ_λ is *not dominant* for $\Delta^+(\mathfrak{k})$ when $\mathbb{F} = \mathbb{H}$ with any p, q, m; when $\mathbb{F} = \mathbb{C}$ with $p \neq 2m$; when $\mathbb{F} = \mathbb{R}$ with $p - q - 2m \geq 3$.

The purpose of this paper is to make a detailed study on Problem$(0.6)'$ when G/H_2 is of the form (0.10), by seeking for useful techniques which are applicable to derived functor modules with singular parameters and aiming at some understanding of the unitary dual. Most of our results of this paper involve a special class of θ-stable parabolic subalgebra: A Levi part is of the form $L \simeq \mathbb{T}^l \times U(p-k, q; \mathbb{F})$. ($l = k$ if $\mathbb{F} = \mathbb{C}, \mathbb{H}$ and $2l = k$ if $\mathbb{F} = \mathbb{R}$.) Our approach here is algebraic and does not depend on the fact that the representations are realized in L^2-functions; even $H = H_1 \times H_2$ will not appear explicitly in the proof. We shall sometimes deal with Problem(0.6)$'$ for less special derived functor modules at the same time in the proof (e.g. §5).

Remark 0.11.

i) The discrete series which were first constructed by Flensted-Jensen for G/H, by Schlichtkrull for G/H_2 have the property that μ_λ is $\Delta^+(\mathfrak{k})$-dominant, which corresponds to the fact that $\mathcal{B}^j_{K^r}(\delta^* \otimes (-\nu))$ contains a measure valued function. We call these discrete series *Flensted-Jensen type*, *Schlichtkrull type*, respectively. Among discrete series of Schlichtkrull type, the representations which have stable Langlands parameters are explicitly stated in Theorem 8.2 $\sim$ 8.4 in [23] (with some minor misprints). Some of them were "new" unitary representations. It is known that Flensted-Jensen type discrete series do not always exhaust all discrete series for $L^2(G/H)$ ([8] §8, [20]). We notice that if $G \supset H \supset H_2$ is of the form (0.10) and if $\mathbb{F} = \mathbb{H}$, the shortage of Flensted-Jensen type for G/H happens only if $p \gg q + 2m$, while that of Schlichtkrull type for G/H_2 happens for any p, q and m. (This is similar if $\mathbb{F} = \mathbb{C}$.)

ii) Clearly, discrete series are unitarizable with respect to their L^2-norm. On the other hand, unitarizability of Zuckerman's derived functor modules has been proven algebraically in the weakly good range [29],[34]. If they are induced from characters, the assumption can be weakened to the weakly fair range ([29], Theorem 7.1). Then the latter stronger version covers the range of discrete series for semisimple symmetric spaces G/H (see List(0.10)). One of the reasons why we reduce our discrete series to cohomologically induced representations from characters (see Remark(0.7)) is to compare with these algebraic results.

It has also been observed by many people that the unitarizability of $\mathcal{R}_{\mathfrak{q}}^{S}(\mathbb{C}_\lambda)$ special parabolic subalgebras $\mathfrak{q}$ still holds outside the fair range with small spectrum λ (see [7] and the references therein). It seems that our unitarizability results in Part(6) of Theorems 1 and 2 contain some new cases, namely, the θ-stable parabolic subalgebras $\mathfrak{q}$ which we treat here are more general.

iii) None of Problem(0.6)$'$ is trivial even if $\mathbb{C}_\lambda$ is in the (weakly) fair range: For example, Problem(0.6) has been recently investigated by Bien, Vogan, Matsuki and Oshima ([2], [21], [33]) for discrete series for G/H (semisimple symmetric spaces). J.F.Adams [1] also deals with some part of Problem(0.6)$'$ in this range when $\mathfrak{q}$ is "holomorphic" in studying unitary highest weight modules.

Now let us explain briefly the methods of this paper to deal with Problem(0.6)$'$.

To find the condition that assures $\mathcal{R}_{\mathfrak{q}}^{S}(\mathbb{C}_\lambda) \neq 0$, there are various techniques such as

i) Coherent continuation based on

a) the τ-invariant (Vogan's U_α calculus) (e.g. [27], [1]),

b) a precise knowledge of the non-vanishing exponents in the asymptotic behavior of spherical functions on semisimple symmetric spaces [22],

c) small parameters out of the fair range.

ii) Generalized Blattner formula.

The analytic approach (i-b) depends on the fact that these modules are realized in eigenspaces on symmetric spaces. This idea was first introduced by Oshima and Matsuki to check the non-vanishing of some part of the discrete series for G/H (unpublished). We do not go into details here. A formulation with a detailed proof in a generalized setting G/H_2 and actual calculations in a special setting may be seen in [13], which was our first approach to this subject.

We take the approach of neither (i-a) nor (i-b) in this paper, because both of them are only partially successful in covering all of the singular parameters in our Theorems 1 to 3 in §2.

The technique (i-c) is new and appears valuable because of its simplicity. The idea is illustrated in §5 for $G = Sp(p,q)$ and more general θ-stable parabolic subalgebras. The point here is to concentrate on the range of most singular and small parameters (which we write $\mathcal{B}$ standing for 'bounded blocks') by forgetting all about dominant conditions. We will find that the picture of the set

$$\{\lambda;\ \mathcal{R}_{\mathfrak{q}}^{S}(\mathbb{C}_\lambda) \neq 0,\ \lambda \text{ is fair with respect to } \mathfrak{q}\}$$

depends heavily on the choice of polarizations of θ-stable parabolic subalgebras with a fixed Levi part.

As for the method (ii), notoriously complicated cancellations of many terms in a generalized Blattner formula have prevented us from an explicit calculation of a general K-type except some few cases such as

- G is small.
- a K-type is special (e.g. with a highest weight μ_λ under (0.8)(c)).
- a θ-stable parabolic subalgebra $\mathfrak{q}$ is special (e.g. quasi abelian (see [7]) or corresponds to unitary highest weight modules) and a K-type is less special.

Another novelty here is to get nice information on *all* K-types in our setting in §4. Let us introduce the idea briefly. Put $\Theta_\lambda := \sum_j (-1)^j \mathcal{R}_{\mathfrak{q}}^{j}(\mathbb{C}_\lambda)$ as a virtual K-module. We stratify $\widehat{K}$ by a *subspace* $\mathfrak{t}$ of a Cartan subalgebra $\mathfrak{t}^c$ of $\mathfrak{k}$, and add all the multiplicities ($\sim$ dimensions) of K-types occurring in Θ_λ belonging to each stratum parametrized by $\delta \in \mathfrak{t}^*$. (If $\mathcal{R}_{\mathfrak{q}}^{S}(\mathbb{C}_\lambda)$ is a so called ladder representation, the parameter space $\mathfrak{t}$ is one dimensional in our definition.) The resulting function $M(\mathfrak{q}, \lambda, \delta)$ (precisely, see §4.1) vanishes for all δ if $\Theta_\lambda = 0$ by definition. Then we present an explicit formula for $M(\mathfrak{q},\lambda,\delta)$ (Propositions in §4.3-5) in terms of the determinant of a matrix whose entries are polynomials of λ and δ, from which we can determine the explicit condition describing whether $\mathcal{R}_{\mathfrak{q}}^S(\mathbb{C}_\lambda)$ vanishes or not under the assumption $\mathcal{R}_{\mathfrak{q}}^j(\mathbb{C}_\lambda) = 0\,(j \neq S)$. The special value of $M(\mathfrak{q},\lambda,\delta)$ for the 'minimal' $\delta := \tilde{\mu_\lambda}$ (see (4.3.4),(4.4.4),(4.5.4) for definition) is also found. It will play an important role in showing pairwise inequivalence among our discrete series in §8. This method here is applicable not only when μ_λ is *not* $\Delta^+(\mathfrak{k})$-dominant but also when λ is small and *out of* the fair range.

We also mention that a related non-vanishing condition (necessity part) has been proven generally by Matsuki [21] for discrete series for semisimple symmetric spaces (i.e. $H_1 = \{e\}$ case). His proof is beautiful but depends on the fact that they are realized in L^2-eigenspaces on symmetric spaces.

To prove the irreducibility of $\mathcal{R}_{\mathfrak{q}}^S(\mathbb{C}_\lambda)$ in the fair range in our settings, we can use the known theory of $\mathcal{D}$-modules except the case $G = Sp(p,q)$ (see Fact(6.2.4)). When $\mathfrak{g} = \mathfrak{sp}(n,\mathbb{C})$, we shall follow Vogan's method in [33], Example 5.10. The main step of the proof is to find a very special direction to which a translation of the twisted differential operators on a generalized flag variety coming from $U(\mathfrak{g})$ behaves reasonably. Although there is *no* new ideas in §6, we shall give a complete proof of this main step (see Theorem(6.3.1), a slight refinement of Theorem 5.11 in [33]) which is formulated more generally than what we need.

Here is an outline of the contents. In §1 we fix some notations and recall related facts. They are restrictive and set the scene for §2. The main results are Theorem 1, 2 and 3 in §2. The rest of the paper is devoted to prove them (sometimes with suitable generalization). Since the parameters outside the fair range are peculiar to our discrete series, our concern will be mainly with $\mathbb{F} = \mathbb{C}$ or $\mathbb{H}$. §3 contains further useful notations and preliminary results on translation functors that we need in the proof. This is essentially standard material included for the benefit of the reader. §4, 5 and 6 are independent of each other and the reader can go directly to any of them. §4 considers the necessary and sufficient condition that $\mathcal{R}_{\mathfrak{q}}^{S}(\mathbb{C}_\lambda) \neq 0$ by a K-type formula. §5 provides another method to prove quickly the sufficient condition for the non-vanishing of Θ_λ in a more generalized situation. (We illustrate this idea when $\mathbb{F} = \mathbb{H}$ case.) §6 gives the proof of irreducibility result in the fair range when $\mathfrak{g} = \mathfrak{sp}(n, \mathbb{C})$ with special parabolic subalgebras. §7 proves some vanishing results $\mathcal{R}_{\mathfrak{q}}^{j}(\mathbb{C}_\lambda) = 0$ $(j \neq S)$ outside the fair range with small spectrum. The proof for pairwise inequivalence of these modules is given in §8.

This paper consists of three lectures which were delivered at Conference on "Eigenfunctions on symmetric spaces and representations of Lie groups" held at RIMS Kyoto on July 1987, Summer school at University of Industrial Technology on August 1987, and Lie groups and representations Seminar at University of Tokyo on January 1989. The results here except §5 were announced in [14] with a sketch of the proof. The author expresses his sincere gratitude to Professor Toshio Oshima for the suggestion of generalizing Schlichtkrull's results which was a motivation of the initial work [13]. Moreover, our results in §4 are inspired by the idea of controlling the multiplicities of 'small' K-types, which I learned from him. It is a pleasure to thank Professor Toshihiko Matsuki who is generously willing to allow me to publish our proof for §5 (quite different from his) here first. I would like to thank Professor Henrik Schlichtkrull who read quite carefully the original manuscript together with Dr. Jesper Bang-Jensen and sent me a list of errors and comments, and Mr. Hiroyuki Ochiai who examined parts of the manuscript. Thanks are also due to the referee for his careful and kind comments. I am also grateful to Professor Mogens Flensted-Jensen for his lectures in Japan in 1986 which stimulated my interest in this field, Professor David Vogan, Dr. Susana Salamanca-Riba and Professor Joseph A. Wolf for their interest in this work, and Dr. Itaru Terada for the instruction of TEX.

§1. Notation

In this section we set up notation. See [28] and [32] for general references. Let H be a connected real linear reductive Lie group, with real Lie algebra $\mathfrak{h}_0$ and complexified Lie algebra $\mathfrak{h}$. Since H is contained in a complex Lie group, we denote by $H_{\mathbb{C}}$ the connected complex Lie subgroup with Lie algebra $\mathfrak{h}$. The center of the universal enveloping algebra $U(\mathfrak{h})$ is written as $\mathfrak{Z}(\mathfrak{h})$. In what follows analogous notation will be applied to Lie groups denoted by other Roman upper case letters without comment.

We will use the standard notation $\mathbb{N}$, $\mathbb{Z}$, $\mathbb{R}$, $\mathbb{C}$ and $\mathbb{H}$. Here $\mathbb{N}$ means the set of non-negative integers and $\mathbb{H}$ means the $\mathbb{R}$-algebra of quarternionic numbers. We denote by $\mathbb{N}_+$ the set of positive integers. For $x \in \mathbb{R}$, we write $[x] := \sup\{n \in \mathbb{Z}; n \leq x\}$, the Gaussian integer of x.

1. θ-stable parabolic subalgebra

Let G be a connected real linear reductive Lie group. Let $K \subset G$ be a maximal compact subgroup and fix a Cartan involution θ so that $\mathfrak{g}_0 = \mathfrak{k}_0 + \mathfrak{p}_0$ is a Cartan decomposition of $\mathfrak{g}_0$. Fix a nondegenerate bilinear form $\langle\, ,\, \rangle$ on $\mathfrak{g}$ invariant under G and θ, which is positive definite on $\mathfrak{p}_0$ and negative definite on $\mathfrak{k}_0$. This form will be restricted to subspaces and transferred to dual vector spaces without change of notation. If the restriction of $\langle\, ,\, \rangle$ to each subspace $\mathfrak{a}$, $\mathfrak{b}$ with $\mathfrak{a} \subset \mathfrak{b}$ is non-degenerate, we look upon

$$\mathfrak{a}^* \subset \mathfrak{b}^* \tag{1.1}$$

through this bilinear form.

Fix an abelian subalgebra $\mathfrak{t}_0$ of $\mathfrak{k}_0$. Define L to be the centralizer of $\mathfrak{t}_0$ in G. Let $\mathfrak{u}$ be the vector space spanned by positive eigenspaces of a fixed generic element of $\sqrt{-1}\mathfrak{t}_0^*$. Then $\mathfrak{q} := \mathfrak{l} + \mathfrak{u}$ gives a Levi decomposition and $\mathfrak{q}$ is called a θ-stable parabolic subalgebra ([28] Definition 5.2.1). Note that L coincides with the normalizer of $\mathfrak{q}$ in G and is a connected reductive Lie group. Clearly $\mathfrak{t}$ is contained in the center of $\mathfrak{l}$, and in this paper we shall usually assume that $\mathfrak{t}$ satisfies

$$\mathfrak{t} \text{ coincides with the center of } \mathfrak{l}. \tag{1.1.2}$$

We use this assumption only for simplifying notation. In most situations we will let

$$S := \dim(\mathfrak{u} \cap \mathfrak{k}) = \frac{1}{2} \dim K/L \cap K. \tag{1.1.3}$$

Write $\mathbb{C}_{2\rho(\mathfrak{u})}$ for the determinant character of L on $\mathfrak{u}$ whose differential is given by

$$2\rho(\mathfrak{u})(X) = \operatorname{trace}(\operatorname{ad}(X)_{|\mathfrak{u}}) \qquad (X \in \mathfrak{l}). \tag{1.1.4}$$

Let $\mathfrak{h}$ be a Cartan subalgebra of $\mathfrak{l}$. Then $\mathfrak{t} \subset \mathfrak{h}$ and $\mathfrak{h}$ is also a Cartan subalgebra of $\mathfrak{g}$. We sometimes view $\rho(\mathfrak{u})$ as an element of $\mathfrak{h}^*$ or of $\mathfrak{t}^*$ under the assumption(1.1.2) according to notation(1.1.1). Let $L^\sim \equiv L^{\sim G}$ be the metapletic two-fold cover of L (Duflo; see [32] Definition 5.7) defined by the square root $\rho(\mathfrak{u})$ of $2\rho(\mathfrak{u})$. (The definition of $L^\sim$ is independent of the particular choice of the nilradical $\mathfrak{u}$). Write ζ for the non-trivial element of the kernel of the covering map $L^\sim \to L$. Then the evaluation of $\rho(\mathfrak{u})$ at ζ is by definition -1. A metapletic representation of $L^\sim$ is the one that is -1 on ζ.

There is a natural bijection between metapletic representations of $L^\sim$ and representations of L by the assignment

$$\tau \mapsto \tau \otimes \mathbb{C}_{-\rho(\mathfrak{u})} \tag{1.1.5}$$

for each metapletic representation τ.

We denote by $\mathbb{C}_\lambda$ the one-dimensional metapletic representation of $L^\sim$ whose differential is given by $\lambda \in \mathfrak{l}^*$. (Then automatically $\lambda \in \mathfrak{t}^*$ in the sense of (1.1.1) under the assumption (1.1.2).) We use the same notation $\mathbb{C}_\lambda$ when it is a character of L or $L^\sim$.

1.2. good range, fair range

Take a Cartan subalgebra $\mathfrak{h}$ of $\mathfrak{g}$ and let $W(\mathfrak{g},\mathfrak{h})$ be the Weyl group of the root system $\Delta(\mathfrak{g},\mathfrak{h})$. We usually assume $\mathfrak{h}_0$ is a fundamental Cartan subalgebra. This means $\mathfrak{h}_0 = \theta\mathfrak{h}_0$ and $\mathfrak{h}_0 = \mathfrak{t}_0^c + \mathfrak{a}_0^c \equiv (\mathfrak{h}_0 \cap \mathfrak{k}_0) + (\mathfrak{h}_0 \cap \mathfrak{p}_0)$ such that $\mathfrak{t}_0^c$ is a Cartan subalgebra of $\mathfrak{k}_0$. A maximal ideal of $\mathfrak{Z}(\mathfrak{g})$ is identified with a $W(\mathfrak{g},\mathfrak{h})$ orbit in $\mathfrak{h}^*$:

$$\operatorname{Hom}_{\mathbb{C}-\text{algebra}}(\mathfrak{Z}(\mathfrak{g}),\mathbb{C}) \simeq \mathfrak{h}^*/_{\sim W(\mathfrak{g},\mathfrak{h})}$$

via the Harish-Chandra isomorphism $\mathfrak{Z}(\mathfrak{g}) \simeq S(\mathfrak{h})^{W(\mathfrak{g},\mathfrak{h})}$ which involves a shift by $\rho(\Delta^+(\mathfrak{g},\mathfrak{h}))$.

We follow the terminologies below from [33] Definition 2.5. Suppose that $\mathfrak{h}$ is contained in $\mathfrak{l}$ in the setting of §1.1.

Definition 1.2.1. Let W be a metapletic $(\mathfrak{l},(L\cap K)^\sim)$-module which has a $\mathfrak{Z}(\mathfrak{l})$-infinitesimal character represented by $\gamma \in \mathfrak{h}^*$. We say that W is *in the good range* if

$$\operatorname{Re}\langle \alpha, \gamma\rangle > 0 \qquad \text{for each } \alpha \in \Delta(\mathfrak{u},\mathfrak{h}). \tag{1.2.2}$$

In this case we also say that γ is in the good range with respect to $\mathfrak{q} \subset \mathfrak{g}$. Clearly, this condition is invariant under the action of the Weyl group $W(\mathfrak{l},\mathfrak{h})$.

Suppose that $[\mathfrak{l},\mathfrak{l}]$ acts by zero on W. We say that W is *in the fair range* if

$$\operatorname{Re}\langle \alpha, \gamma_{|\mathfrak{t}}\rangle > 0 \qquad \text{for each } \alpha \in \Delta(\mathfrak{u},\mathfrak{h}) \tag{1.2.2}$$

(notation §1.1). W (or γ) is called *weakly good* (respectively *weakly fair*) if the weak inequalities ($\geq$) hold.

One should notice that good implies fair if $[\mathfrak{l}, \mathfrak{l}]$ acts by zero on W. Conversely, there exists in general a strip of the fair range that is not in the good range. Roughly, the size of this strip corresponds to that of a Levi part of $\mathfrak{q}$ and it is empty when $\mathfrak{q}$ is a Borel subalgebra.

1.3. cohomological parabolic induction

As an algebraic analogue of Dolbeault cohomology on a homogeneous complex manifold G/L, Zuckerman introduced the cohomological parabolic induction (we follow [32] Definition 6.20)

$$\mathcal{R}_{\mathfrak{q}}^j \equiv \left(\mathcal{R}_{\mathfrak{q}}^{\mathfrak{g}}\right)^j \quad (j \in \mathbb{N}),$$

which is a covariant functor from the category of metapletic $(\mathfrak{l}, (L \cap K)^{\sim})$-modules to the category of $(\mathfrak{g}, K)$-modules.

With notation as before, fix a Cartan subalgebra $\mathfrak{h} \subset \mathfrak{l}$. The definition here differs from [28] Definition 6.3.1 only by a ρ-shift. In our normalization, if a metapletic $(\mathfrak{l}, (L \cap K)^{\sim})$-module W has $\mathfrak{Z}(\mathfrak{l})$-infinitesimal character $\gamma \in \mathfrak{h}^*$, then $\mathcal{R}_{\mathfrak{q}}^j(W)$ has $\mathfrak{Z}(\mathfrak{g})$-infinitesimal character γ in the Harish-Chandra parametrization.

1.4. results from Zuckerman and Vogan

Retain notations as in §1.1-2. The following theorem is due to Zuckerman and Vogan (see [28],[32],[35]).

Fact 1.4.1. *In the setting of §1.1, suppose that W is a metapletic $(\mathfrak{l}, (L \cap K)^{\sim})$-module.*

1) *Assume that W is weakly good.*

a) $\mathcal{R}_{\mathfrak{q}}^j(W) = 0$ *for all* $j \neq S$.

b) $\mathcal{R}_{\mathfrak{q}}^{S}(W)$ *is an irreducible* $(\mathfrak{g}, K)$*-module or zero if* W *is irreducible.*

c) $\mathcal{R}_{\mathfrak{q}}^{S}(W)$ *is unitarizable if* W *is unitary.*

2) *If* $W(\neq 0)$ *is good, then* $\mathcal{R}_{\mathfrak{q}}^{S}(W) \neq 0$.

3) *If* W *is good and if* $\mathcal{R}_{\mathfrak{q}}^{S}(W)$ *is unitary, then* W *is unitarizable.*

When W is fair, it is known that some of the properties in the above theorem fail:

Fact 1.4.2. *Retain notations as in* Fact(1.4.1) *and assume that* $[\mathfrak{l}, \mathfrak{l}]$ *acts by zero on* W.

1) *Assume that* W *is weakly fair.*

a) **(true)** $\mathcal{R}_{\mathfrak{q}}^{j}(W) = 0$ *for all* $j \neq S$.

b) **(false)** $\mathcal{R}_{\mathfrak{q}}^{S}(W)$ *is an irreducible* $(\mathfrak{g}, K)$*-module or zero.*

c) **(true)** $\mathcal{R}_{\mathfrak{q}}^{S}(W)$ *is unitarizable if* W *is unitary.*

2) **(false)** *If* W *is fair, then* $\mathcal{R}_{\mathfrak{q}}^{S}(W) \neq 0$.

Part(1)(a),(c) is due to Vogan [29], Theorem7.1. A counter example for (1)(b) is given in [31] when $\mathfrak{g}$ is of type C_4. As for (2), see also §4 and §5.

Note that if W is fair $\mathcal{R}_{\mathfrak{q}}^{j}(W)$ does not necessarily have a regular $\mathfrak{Z}(\mathfrak{g})$-infinitesimal character and does not always have nice behavior under translation.

1.5. results from Harish-Chandra and Oshima-Matsuki

We do not use directly the results cited here in later sections. However, they will help us to understand what we are doing in this paper. Retain notations in §1.1-3. Cohomologically induced representations from θ-stable parabolic subalgebras are convenient in describing 'discrete series':

Fact 1.5.1 (Harish-Chandra). *Any discrete series for G is of the form $\mathcal{R}_{\mathfrak{q}}^S(\mathbb{C}_\lambda)$ with $\mathfrak{t}$ a Cartan subalgebra of $\mathfrak{g}$ contained in $\mathfrak{k}$ and with $\mathbb{C}_\lambda$ in the good range. (In this case $\mathfrak{q}$ is a Borel subalgebra of $\mathfrak{g}$.)*

Fact 1.5.2 (Oshima-Matsuki [20], cf. [9], [33] Definition(2.8)). *Let σ be an involution of G commuting with the Cartan involution θ, and H be an open subgroup of the fixed points of σ. Then any discrete series for a semisimple symmetric space G/H is of the form $\mathcal{R}_{\mathfrak{q}}^S(\mathbb{C}_\lambda)$ with $\mathfrak{t}$ a Cartan subalgebra of $\mathfrak{k}$ contained in $\{X \in \mathfrak{k};\ \sigma(X) = -X\}$ and with $\mathbb{C}_\lambda$ in the fair range.*

Remark 1.5.3. Several remarks are in order. First, the existence of such an abelian subspace $\mathfrak{t}$ is equivalent to the famous rank condition $\operatorname{rank} G = \operatorname{rank} K$ (respectively, $\operatorname{rank} G/H = \operatorname{rank} K/H \cap K$), which is known to be the necessary and sufficient condition for the existence of discrete series. Second, Oshima and Matsuki do not use derived functor modules directly as alluded to in the Introduction. Third, we have mentioned neither the necessary evenness condition on λ (e.g. (2.7.4)(c)) nor the condition for $\mathcal{R}_{\mathfrak{q}}^S(\mathbb{C}_\lambda) \neq 0$. Finally, there exist now algebraic proofs of the unitarizability of these modules (Fact(1.4.2)).

§2. Main results

In this section, we state our main theorems.

2. $G = Sp(p,q)$

Let $G = Sp(p,q)$ where $p,\ q \geq 1$ and let θ be a Cartan involution of $\mathfrak{g}$ corresponding to the maximal compact subgroup $K = Sp(p) \times Sp(q)$ naturally embedded in a matrix group G.

Let $\mathfrak{h}_0$ be a fundamental Cartan subalgebra of $\mathfrak{g}_0$; then $\mathfrak{h}_0$ is contained in $\mathfrak{k}_0$. Choose coordinates $\{f_i\,;\ 1 \leq i \leq p+q\}$ on $\mathfrak{h}^*$ such that the root systems of $\mathfrak{g}$ and $\mathfrak{k}$ for $\mathfrak{h}$ are given by,

$$\Delta(\mathfrak{g},\mathfrak{h}) = \{\pm(f_i \pm f_j), \pm 2f_l\,;\ 1 \leq i < j \leq p+q, 1 \leq l \leq p+q\},$$

$$\begin{aligned}\Delta(\mathfrak{k},\mathfrak{h}) = \{&\pm(f_i \pm f_j)\,;\ 1 \leq i < j \leq p \text{ or } p+1 \leq i < j \leq q\}\\ &\cup \{\pm 2f_l\,;\ 1 \leq l \leq p+q\}.\end{aligned}$$

Let $\{H_i\} \subset \mathfrak{h}$ be the dual basis for $\{f_i\} \subset \mathfrak{h}^*$. Fix an integer r such that $1 \leq r \leq p$. Set

$$\mathfrak{t} := \sum_{j=1}^{r} \mathbb{C}H_j \quad (\subset \mathfrak{h}).$$

Let $\mathfrak{l}$ be the centralizer of $\mathfrak{t}$ in $\mathfrak{g}$, and L be the centralizer of $\mathfrak{t}$ in G. Then L is θ-stable in G, isomorphic to $\mathbb{T}^r \times Sp(p-r,q)$ with complexified Lie algebra $\mathfrak{l}$.

Let $\mathfrak{q} \equiv \mathfrak{q}(r) = \mathfrak{l} + \mathfrak{u}$ (Levi decomposition) be a θ-stable parabolic subalgebra of $\mathfrak{g}$, whose nilpotent radical $\mathfrak{u}$ is defined by the following roots for $\mathfrak{h}$.

$$\Delta(\mathfrak{u},\mathfrak{h}) := \{f_i \pm f_j, 2f_l\,;\ 1 \leq i \leq r, i < j \leq p+q, 1 \leq l \leq r\}.$$

Then $\rho(\mathfrak{u}) = \sum_{j=1}^{r}(p+q+1-j)f_j$, $S \equiv \dim_{\mathbb{C}}(\mathfrak{u} \cap \mathfrak{k}) = r(2p-r)$.

Set $Q := p+q-r(>0)$. Let

$$\lambda := \sum_{i=1}^{r} \lambda_i f_i \quad (\lambda_i \in \mathbb{C})$$

be an element of $\mathfrak{t}^*$, which will be sometimes regarded as an element of $\mathfrak{h}^*$.

2.2. main theorem for $G = Sp(p,q)$

Theorem 1. *Retain notations in §2.1.*

0) *Any θ-stable parabolic subalgebra of $\mathfrak{g}$ with Levi part $\mathfrak{l}$ is conjugate to $\mathfrak{q} = \mathfrak{l} + \mathfrak{u}$ by an element of K.*

1) *$\mathbb{C}_\lambda$ lifts to a metapletic $(\mathfrak{l}, (L \cap K)^{\sim})$-module if and only if $\lambda_i \in \mathbb{Z}$ for all i $(1 \le i \le r)$.*

From Part (2) to (6), we always assume

$$\lambda_i \in \mathbb{Z} \quad (1 \le i \le r). \tag{2.2.1}$$

2) *If λ satisfies*

$$\lambda_1 \ge \lambda_2 \ge \cdots \ge \lambda_{r-1} \ge |\lambda_r| \quad \text{and} \quad \lambda_r \ge -Q, \tag{2.2.2}$$

then $\mathcal{R}_{\mathfrak{q}}^{S-j}(\mathbb{C}_\lambda) = 0$, for any $j \ne 0$.

3) *Under the assumption (2.2.2), $\mathcal{R}_{\mathfrak{q}}^{S}(\mathbb{C}_\lambda) \ne 0$ if and only if*

$$\begin{cases} \lambda_1 \ge \lambda_2 \ge \cdots \ge \lambda_{r-1} \ge |\lambda_r| \quad \text{and} \quad \lambda_r \ge -Q, \\ \lambda_i \ne \lambda_j \qquad (i \ne j) \\ \lambda_{r-2q} \ge Q+1 \qquad (\text{when } r > 2q). \end{cases} \tag{2.2.3}$$

Here we impose the third condition only when $r > 2q$.

If (2.2.3) is satisfied, then $\mathcal{R}_{\mathfrak{q}}^S(\mathbb{C}_\lambda)$ has $\mathfrak{Z}(\mathfrak{g})$-infinitesimal character

$$(\lambda_1, \ldots, \lambda_r, Q, Q-1, \ldots, 1) \in \mathfrak{h}^*$$

in the Harish-Chandra parametrization.

4) *The set $\{\mathcal{R}_{\mathfrak{q}}^S(\mathbb{C}_\lambda);\ \lambda$ satisfies (2.2.3)$\}$ consists of non-zero $(\mathfrak{g}, K)$-modules which are pairwise inequivalent.*

5) *If λ satisfies (2.2.3) and $\lambda_r > 0$, then $\mathcal{R}_{\mathfrak{q}}^S(\mathbb{C}_\lambda)$ is a non-zero irreducible $(\mathfrak{g}, K)$-module.*

6) *Assume that r is an even number. Set $r = 2m$ $(0 < 2m \leq p)$. If λ satisfies (2.2.3) and $\lambda_{r-1} + \lambda_r > 0$, then there is an injective $(\mathfrak{g}, K)$-homomorphism*

$$\mathcal{R}_{\mathfrak{q}}^S(\mathbb{C}_\lambda) \longrightarrow L^2(Sp(p,q)/Sp(p-m,q))$$

into discrete series for an indefinite quarternionic Stiefel manifold $G/H_2 = Sp(p,q)/Sp(p-m,q)$. In particular, the corresponding derived functor modules are non-zero unitarizable.

Remark 2.2.4. We can derive the unitarizability of $\mathcal{R}_{\mathfrak{q}}^S(\mathbb{C}_\lambda)$ from Part(6) in the above theorem under the assumption (2.2.3). In fact, the case $\lambda_{r-1} + \lambda_r = 0$ with $r = 2m$ corresponds to "limit of discrete series" for $Sp(p,q)/Sp(p-m,q)$ and is still unitarizable. As for an odd integer r, we can deduce the unitarizability from the even integer $r+1$ case. Actually, we choose sufficiently large λ_1 and make use of Fact(1.4.1)(3), Lemma(3.2.1)(2).

Remark 2.2.5. If we apply our result to the case where $r = 1$ under (2.2.3), we have the known result on the unitarizability, the non-vanishing, and an explicit K-type formula (see also §4.3) of $\mathcal{R}_{\mathfrak{q}}^S(\mathbb{C}_\lambda)$ due to Enright, Parthasarathy, Wallach, and Wolf (see [7], §9). In this case ($r = 1$), $\mathfrak{q}$ is quasi abelian and $\mathcal{R}_{\mathfrak{q}}^S(\mathbb{C}_\lambda)$ is a ladder representation in their terminology. Our assumption(2.2.3) is nothing but $\lambda_1 \geq -Q$ when $r = 1$, which is equivalent to "$z < 0$" after a careful comparison of ρ-shift (This range is referred as "especially curious" in [7], p126).

2.3. $G = U(p,q)$

Let $G = U(p,q)$ where p, $q \geq 1$ and let θ be a Cartan involution corresponding to the maximal compact subgroup $K = U(p) \times U(q)$ naturally embedded in a matrix group G.

Let $\mathfrak{h}_0$ be a fundamental Cartan subalgebra of $\mathfrak{g}_0$; then $\mathfrak{h}_0$ is contained in $\mathfrak{k}_0$. Choose coordinates $\{f_i\,;\, 1 \leq i \leq p+q\}$ on $\mathfrak{h}^*$ such that the root systems of $\mathfrak{g}$ and $\mathfrak{k}$ for $\mathfrak{h}$ are given by

$$\Delta(\mathfrak{g},\mathfrak{h}) = \{\pm(f_i - f_j)\,;\, 1 \leq i < j \leq p+q\},$$

$$\Delta(\mathfrak{k},\mathfrak{h}) = \{\pm(f_i - f_j)\,;\, 1 \leq i < j \leq p \text{ or } p+1 \leq i < j \leq p+q\}.$$

Let $\{H_i\} \subset \mathfrak{h}$ be the dual basis of $\{f_i\} \subset \mathfrak{h}^*$. Fix two nonnegative integers r and s such that $1 \leq r+s \leq p$. Set

$$\mathfrak{t} := \sum_{i=1}^{r+s} \mathbb{C}H_i + \mathbb{C}\sum_{i=1}^{p+q} H_i \quad (\subset \mathfrak{h}).$$

Then the centralizer L of $\mathfrak{t}$ in G is isomorphic to $\mathbb{T}^{r+s} \times U(p-r-s,q)$.

Let $\mathfrak{q} \equiv \mathfrak{q}(r,s) = \mathfrak{l} + \mathfrak{u}$ be a θ-stable parabolic subalgebra of $\mathfrak{g}$ with nilpotent radical $\mathfrak{u}$ given by,

$$\begin{aligned}\Delta(\mathfrak{u},\mathfrak{h}) := &\{f_i - f_j\,;\, 1 \leq i < j \leq r+s\}\\ &\cup\ \{f_i - f_j\,;\, 1 \leq i \leq r,\ r+s+1 \leq j \leq p+q\}\\ &\cup\ \{-f_i + f_j\,;\, r+1 \leq i \leq r+s,\ r+s+1 \leq j \leq p+q\}.\end{aligned}$$

Then $S = \dfrac{1}{2}(r+s)(2p-r-s-1)$. Set $Q := \dfrac{1}{2}(p+q-r-s-1)\,(\geq 0)$.

Finally, let

$$\lambda := \sum_{i=1}^{r+s} \lambda_i f_i + \frac{-r+s}{2} \sum_{i=1}^{p+q} f_i \in \mathfrak{t}^* \quad (\lambda_i \in \mathbb{C}).$$

2.4. main theorem for $G = U(p,q)$

Theorem 2. *Retain notations in §2.3 and fix r, s.*

0) *There are $r+s+1$ K-conjugacy classes of θ-stable parabolic subalgebras of $\mathfrak{g}$ with Levi part $\mathfrak{l}$. A complete system of representatives is given by*

$$\{\mathfrak{q}(a,b)\,;\, a,b \in \mathbb{N}, a+b = r+s\}.$$

1) *$\mathbb{C}_\lambda$ lifts to a metapletic $(\mathfrak{l},(L\cap K)^\sim)$-module if and only if $\lambda_i \in \mathbb{Z}+Q$ for all $i\,(1 \le i \le r+s)$.*

From Part (2) to (6), we assume

$$\lambda_i \in \mathbb{Z}+Q \qquad (1 \le i \le r+s). \tag{2.4.1}$$

2) *If λ satisfies*

$$\begin{cases} \lambda_1 \ge \lambda_2 \ge \cdots \ge \lambda_r \ge \lambda_{r+1} \ge \cdots \ge \lambda_{r+s}, \\ \lambda_r \ge -Q & \text{when } r > 0, \\ \lambda_{r+1} \le Q & \text{when } s > 0, \end{cases} \tag{2.4.2}$$

then $\mathcal{R}_{\mathfrak{q}}^{S-j}(\mathbb{C}_\lambda) = 0$ for any $j \neq 0$.

3) *Under the assumption* (2.4.2), $\mathcal{R}_{\mathfrak{q}}^{S}(\mathbb{C}_\lambda) \neq 0$ *if and only if*

$$(2.4.3) \quad \begin{cases} \lambda_1 > \lambda_2 > \cdots > \lambda_r \geq \lambda_{r+1} > \lambda_{r+2} > \cdots > \lambda_{r+s}, \\ \lambda_{r-q} \geq Q+1 & \text{when } r > q, \\ \lambda_r \geq -Q & \text{when } r > 0, \\ \lambda_{r+q+1} \leq -Q-1 & \text{when } s > q, \\ \lambda_{r+1} \leq Q & \text{when } s > 0. \end{cases}$$

If (2.4.3) *is satisfied, then* $\mathcal{R}_{\mathfrak{q}}^{S}(\mathbb{C}_\lambda)$ *has* $\mathfrak{Z}(\mathfrak{g})$*-infinitesimal character*

$$(\lambda_1, \ldots, \lambda_{r+s}, Q, Q-1, \ldots, -Q) + \frac{-r+s}{2}(1, 1, \ldots, 1) \in \mathfrak{h}^*.$$

4) *The set* $\{\mathcal{R}_{\mathfrak{q}}^{S}(\mathbb{C}_\lambda)\, ;\, \lambda$ *satisfies* $(2.4.3)\}$ *consists of non-zero* $(\mathfrak{g}, K)$*-modules which are pairwise inequivalent.*

5) *If* λ *satisfies* (2.4.3) *and* $\lambda_r \geq 0 \geq \lambda_{r+s}$, *then* $\mathcal{R}_{\mathfrak{q}}^{S}(\mathbb{C}_\lambda)$ *is a non-zero irreducible* $(\mathfrak{g}, K)$*-module.*

6) *Assume that* $r = s$. *Set* $m := r\,(= s)$. *If* λ *satisfies* (2.4.3) *and* $\lambda_r > \lambda_{r+1}$, *then there is an injective* $(\mathfrak{g}, K)$*-homomorphism*

$$\mathcal{R}_{\mathfrak{q}}^{S}(\mathbb{C}_\lambda) \longrightarrow L^2(U(p,q)/U(p-m,q))$$

into discrete series for an indefinite complex Stiefel manifold $U(p,q)/U(p-m,q)$. *In particular, the corresponding derived functor modules are non-zero unitarizable.*

Remark 2.4.4. The unitarizability of $\mathcal{R}_{\mathfrak{q}}^{S}(\mathbb{C}_\lambda)$ under the assumption (2.4.3) (without assuming $r = s$) is derived from Part(6) in the above theorem by reasoning similar to Remark(2.2.4)

Remark 2.4.5. Our parabolic subalgebra $\mathfrak{q} \equiv \mathfrak{q}(r,s)$ is holomorphic ([1], Definition 1.6) iff $r=0$ or $s=0$. $\mathfrak{q} \equiv \mathfrak{q}(r,s)$ is quasi abelian ([7], p.109) iff $(r,s)=(1,0),(1,1)$ or $(0,1)$. Another algebraic approach of some part of our results may be seen in the above cases (cf. Part(ii) and (iii) in Remark(0.11)).

2.5. $G = SO_0(p,q)$

Let $G = SO_0(p,q)$ where $p,\ q \geq 1$ and let θ be a Cartan involution corresponding to the maximal compact subgroup $K = SO(p) \times SO(q)$ naturally embedded in a matrix group G.

Let $\mathfrak{h}_0 = (\mathfrak{h}_0 \cap \mathfrak{k}_0) + (\mathfrak{h}_0 \cap \mathfrak{p}_0) \equiv \mathfrak{t}_0^c + \mathfrak{a}_0^c$ be a fundamental Cartan subalgebra of $\mathfrak{g}$. If both p and q are odd, then $\dim_{\mathbb{C}} \mathfrak{a}^c = 1$, otherwise $\dim_{\mathbb{C}} \mathfrak{a}^c = 0$ and $\mathfrak{h} = \mathfrak{t}^c$. Choose coordinates $\{f_i\,;\, 1 \leq i \leq [\frac{p+q}{2}]\}$ on $\mathfrak{h}^*$ such that the root systems of $\mathfrak{g}$ and $\mathfrak{k}$ are given by

$$\begin{aligned}\Delta(\mathfrak{g},\mathfrak{h}) =& \{\pm(f_i \pm f_j)\,;\, 1 \leq i < j \leq [\frac{p+q}{2}]\}\\ &\cup \left(\{\pm f_l\,;\, 1 \leq l \leq [\frac{p+q}{2}]\} \ (p+q\text{:odd})\right)\end{aligned}$$

$$\begin{aligned}\Delta(\mathfrak{k},\mathfrak{t}^c) =& \{\pm(f_i \pm f_j)\,;\, 1 \leq i < j \leq [\frac{p}{2}]\}\\ &\cup \{\pm(f_i \pm f_j)\,;\, [\frac{p+q}{2}] - [\frac{q}{2}] + 1 \leq i < j \leq [\frac{p+q}{2}]\}\\ &\cup \left(\{\pm f_l\,;\, 1 \leq l \leq [\frac{p}{2}]\} \ (p\text{:odd})\right)\\ &\cup \left(\{\pm f_l\,;\, [\frac{p+q}{2}] - [\frac{q}{2}] + 1 \leq l \leq [\frac{p+q}{2}]\} \ (q\text{:odd})\right),\end{aligned}$$

respectively. Let $\{H_i\} \subset \mathfrak{h}$ be the dual basis for $\{f_i\} \subset \mathfrak{h}^*$. Fix an integer r such that $1 \leq r \leq [\frac{p}{2}]$. Set $\mathfrak{t} := \sum_{j=1}^{r} \mathbb{C}H_j$ $(\subset \mathfrak{t}^c \subset \mathfrak{h})$. Then the centralizer L of $\mathfrak{t}$ in G is isomorphic to $\mathbb{T}^r \times SO_0(p-2r,q)$. Define two elements of $\mathfrak{t}^*$ $(\subset (\mathfrak{t}^c)^* \subset \mathfrak{h}^*)$ by,

$$\mu := \sum_{i=1}^{r} (\frac{p+q}{2} - i) f_i, \quad \mu' := \mu - (p+q-2r) f_r \in \mathfrak{t}^*.$$

Let $\mathfrak{q} = \mathfrak{l} + \mathfrak{u}$ (resp. $\mathfrak{q}' := \mathfrak{l} + \mathfrak{u}'$) be a θ-stable parabolic subalgebra of $\mathfrak{g}$ with nilpotent radical $\mathfrak{u}$ (resp. $\mathfrak{u}'$) given by,

$$\Delta(\mathfrak{u}, \mathfrak{h}) := \{\alpha \in \Delta(\mathfrak{g}, \mathfrak{h})\, ;\, \langle \alpha, \mu \rangle > 0\},$$

$$\Delta(\mathfrak{u}', \mathfrak{h}) := \{\alpha \in \Delta(\mathfrak{g}, \mathfrak{h})\, ;\, \langle \alpha, \mu' \rangle > 0\}.$$

Then $\rho(\mathfrak{u}) = \mu$ (resp. $\rho(\mathfrak{u}') = \mu'$) when restricted to $\mathfrak{h}$ and $S = r(p - r - 1)$. Set $Q := \dfrac{p+q}{2} - r - 1 (\geq -\dfrac{1}{2})$. Finally, let

$$\lambda := \sum_{i=1}^{r} \lambda_i f_i, \quad \lambda' := \lambda - 2\lambda_r f_r \in \mathfrak{t}^* \quad (\lambda_i \in \mathbb{C}).$$

2.6. main theorem for $G = SO_0(p,q)$

Theorem 3. *Retain notations in §2.5.*

0) *A complete system of representatives of the K-conjugacy classes of θ-stable parabolic subalgebras of $\mathfrak{g}$ with Levi part $\mathfrak{l}$ is given by,*

   a) $\{\mathfrak{q}\}$ *when $p \neq 2r$,*

   b) $\{\mathfrak{q}\}, \{\mathfrak{q}'\}$ *when $p = 2r$.*

1) *The following three conditions on λ are equivalent:*

   a) *$\mathbb{C}_\lambda$ lifts to a metapletic $(\mathfrak{l}, (L \cap K)^\sim)$-module.*

   b) *$\mathbb{C}_{\lambda'}$ lifts to a metapletic $(\mathfrak{l}, (L \cap K)^\sim)$-module.*

   c) *$\lambda_i \in \mathbb{Z} + Q$ for all i $(1 \leq i \leq r)$.*

   From Part (2) to (6), we always assume

$$\lambda_i \in \mathbb{Z} + Q \quad (1 \leq i \leq r). \tag{2.6.1}$$

2) *If* λ *satisfies*

$$\lambda_1 \geq \lambda_2 \geq \cdots \geq \lambda_r \geq 0, \tag{2.6.2}$$

then $\mathcal{R}_{\mathfrak{q}}^{S-j}(\mathbb{C}_\lambda) = \mathcal{R}_{\mathfrak{q}'}^{S-j}(\mathbb{C}_{\lambda'}) = 0$ *for any* $j \neq 0$.

3) *Under the assumption* (2.6.2), *the following three conditions are equivalent.*

a) $\mathcal{R}_{\mathfrak{q}}^{S}(\mathbb{C}_\lambda) \neq 0$.

b) $\mathcal{R}_{\mathfrak{q}'}^{S}(\mathbb{C}_{\lambda'}) \neq 0$.

c) λ *satisfies*

$$\begin{cases} \lambda_1 > \lambda_2 > \cdots > \lambda_r \geq 0, \\ \lambda_{r-q} \geq Q+1 \qquad (\text{when } r > q). \end{cases} \tag{2.6.3}$$

In this case, $\mathcal{R}_{\mathfrak{q}}^{S}(\mathbb{C}_\lambda)$ *and* $\mathcal{R}_{\mathfrak{q}'}^{S}(\mathbb{C}_{\lambda'})$ *have the same* $\mathfrak{Z}(\mathfrak{g})$*-infinitesimal character*

$$(\lambda_1, \ldots, \lambda_r, Q, Q-1, \ldots, Q-[Q]) \in \mathfrak{h}^*.$$

4) *When* $p \neq 2r$, $\mathcal{R}_{\mathfrak{q}}^{S}(\mathbb{C}_\lambda) \simeq \mathcal{R}_{\mathfrak{q}'}^{S}(\mathbb{C}_{\lambda'})$ *and the set*

$$\{\mathcal{R}_{\mathfrak{q}}^{S}(\mathbb{C}_\lambda);\ \lambda \text{ satisfies } (2.6.3)\}$$

consists of pairwise inequivalent $(\mathfrak{g}, K)$*-modules.*

When $p = 2r$, *the set*

$$\{\mathcal{R}_{\mathfrak{q}}^{S}(\mathbb{C}_\lambda),\ \mathcal{R}_{\mathfrak{q}'}^{S}(\mathbb{C}_{\lambda'});\ \lambda \text{ satisfies } (2.6.3)\}$$

consists of pairwise inequivalent $(\mathfrak{g}, K)$*-modules.*

5) *If* λ *satisfies* (2.6.3), *then the* $(\mathfrak{g}, K)$*-modules* $\mathcal{R}_{\mathfrak{q}}^{S}(\mathbb{C}_\lambda)$ *and* $\mathcal{R}_{\mathfrak{q}'}^{S}(\mathbb{C}_{\lambda'})$ *are non-zero and irreducible.*

6) *If λ satisfies (2.6.3) and $\lambda_r > 0$, then there is an injective $(\mathfrak{g}, K)$-homomorphism*

$$\mathcal{R}_{\mathfrak{q}}^{S}(\mathbb{C}_\lambda) \longrightarrow L^2(SO_0(p,q)/SO_0(p-r,q)), \text{ when } p \neq 2r,$$

$$\mathcal{R}_{\mathfrak{q}}^{S}(\mathbb{C}_\lambda) \oplus \mathcal{R}_{\mathfrak{q}'}^{S}(\mathbb{C}_{\lambda'}) \longrightarrow L^2(SO_0(p,q)/SO_0(p-r,q)), \text{ when } p = 2r,$$

into discrete series for an indefinite real Stiefel manifold $SO_0(p,q)/SO_0(p-r,q)$.

Remark 2.6.4. As opposed to the quarternionic case or the complex case, our result for real pseudo-orthogonal groups, on the unitarizability of derived functor modules, is completely contained in general theory because the parameter is always in the fair range.

2.7. list and figures of various conditions on parameters

For reference below we list various conditions on λ explicitly. We denote by (a) the case where $G = Sp(p,q)$, by (b) the case where $G = U(p,q)$, and by (c) the case where $G = SO(p,q)$ respectively.

(2.7.0)(a) $$Q = p + q - r,$$

(2.7.0)(b) $$Q = \frac{1}{2}(p + q - r - s - 1),$$

(2.7.0)(c) $$Q = \frac{1}{2}(p + q) - r - 1.$$

The condition that $\mathbb{C}_\lambda$ is in the weakly fair range amounts to

(2.7.1)(a) $$\lambda_1 \geq \cdots \geq \lambda_r \geq 0,$$

(2.7.1)(b) $$\lambda_1 \geq \cdots \geq \lambda_r \geq 0 \geq \lambda_{r+1} \geq \cdots \geq \lambda_{r+s},$$

(2.7.1)(c) $$\lambda_1 \geq \cdots \geq \lambda_r \geq 0.$$

The condition that $\mathbb{C}_\lambda$ is in the weakly good range amounts to

(2.7.2)(a) $\lambda_1 \geq \cdots \geq \lambda_r \geq Q,$

(2.7.2)(b) $\lambda_1 \geq \cdots \geq \lambda_r \geq Q$ and $-Q \geq \lambda_{r+1} \geq \cdots \geq \lambda_{r+s},$

(2.7.2)(c) $\lambda_1 \geq \cdots \geq \lambda_r \geq \max(0, Q).$

The condition that μ_λ is dominant for $\Delta^+(\mathfrak{k})$ amounts to

(2.7.3)(a) $\lambda_1 > \cdots > \lambda_r \geq p - q - r + 1,$

(2.7.3)(b) $$\begin{cases} \lambda_1 > \cdots > \lambda_r \geq \dfrac{p-q-r-s+1}{2}, \\ \lambda_{r+1} > \cdots > \lambda_{r+s} \geq -\dfrac{p-q-r-s+1}{2}, \end{cases}$$

(2.7.3)(c) $\lambda_1 > \cdots > \lambda_r \geq \dfrac{p-q-2r}{2}.$

Let $r = 2m$ in (a), $r = s = m$ in (b) and $r = m$ in (c), respectively.
The condition that $\mathcal{R}_\mathfrak{q}^S(\mathbb{C}_\lambda)$ is discrete series for a semisimple symmetric space $G/H = U(p,q;\mathbb{F})/U(m;\mathbb{F}) \times U(p-m,q;\mathbb{F})$ amounts to the following condition (conversely, discrete series for G/H are exhausted): $\lambda_i \in \mathbb{Z} + Q$, $\mathbb{C}_\lambda$ is in the fair range, and

(2.7.4)(a) $\lambda_{2i-1} = \lambda_{2i} + 1 \qquad (1 \leq i \leq m),$

(2.7.4)(b) $\lambda_i = -\lambda_{2r+1-i} \qquad (1 \leq i \leq m),$

(2.7.4)(c) $\lambda_{i+1} - \lambda_i \in 2\mathbb{Z} + 1 \quad (1 \leq i \leq m-1).$

Example 2.7.5. Retain notation in Theorem 2 (6) and choose $r = s = m = 1$. Let $G = U(p,q) \supset H = U(1) \times U(p-1,q) \supset H_2 = U(p-1,q)$ with $p+q = 9$, $2 \le p \le 8$. In particular, $\operatorname{rank} G/H = 1$.

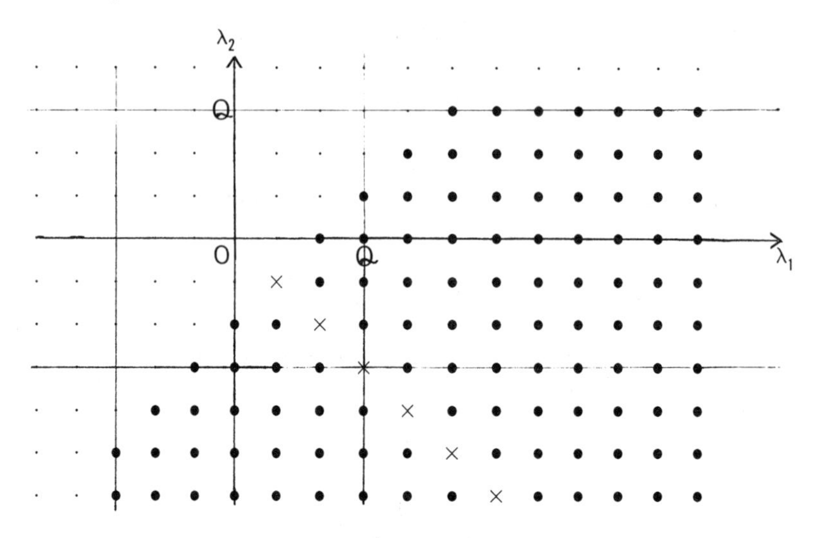

Figure 2.7.6. $(Q = \frac{1}{2}(p+q-3) = 3)$

$\bullet$: $\mathcal{R}_{\mathfrak{q}}^S(\mathbb{C}_\lambda)\,(\neq 0)$ is discrete series for G/H_2 given in Theorem 2 (6);
$(\lambda_1, \lambda_2) \in \mathbb{Z}^2$, $\lambda_1 > \lambda_2$, $-3 \le \lambda_1$ and $\lambda_2 \le 3$.

$\times$: $\mathcal{R}_{\mathfrak{q}}^S(\mathbb{C}_\lambda)\,(\neq 0)$ is discrete series for G/H; $\lambda_1 = -\lambda_2 \in \mathbb{N}_+$.

μ_λ is $\Delta^+(\mathfrak{k})$-dominant (Schlichtkrull type) $\Leftrightarrow \lambda_1 > 2, p-5 \le \lambda_1$ and $\lambda_2 \le -p+5$,

$\mathbb{C}_\lambda$ is weakly fair (cf. Fact(1.4.2)) $\Leftrightarrow \lambda_1 \ge 0 \ge \lambda_2$,

$\mathbb{C}_\lambda$ is good (cf. Fact(1.4.1)) $\Leftrightarrow \lambda_1 > 3 > -3 > \lambda_2$,

$\mathcal{R}_{\mathfrak{q}}^S(\mathbb{C}_\lambda)$ has singular infinitesimal character $\Leftrightarrow \lambda_1(\text{or } \lambda_2) \in \{0, \pm 1, \pm 2, \pm 3\}$.

The figure for $G/H_2 = U(p,q)/U(p-1,q)$ $(2 \le p,\ 1 \le q)$ is similar via the usual identification $\sqrt{-1}\mathfrak{t}_0^* \simeq \mathbb{R}^2$. Roughly, $\lambda_1 - \lambda_2$ (resp. $\lambda_1 + \lambda_2$) determines the asymptotic behavior (resp. the right action of $H_1 = U(1)$) of the functions on G/H_2 contained in a $(\mathfrak{g}, K)$-module $\mathcal{R}_{\mathfrak{q}}^S(\mathbb{C}_\lambda)$.

Example 2.7.9. Retain notation in Theorem 1 (6) and choose $r = 2m = 2$. Let $G = Sp(p,q) \supset H = Sp(1) \times Sp(p-1,q) \supset H_2 = Sp(p-1,q)$ with $p+q=5$, $2 \le p \le 4$. In particular, $\operatorname{rank} G/H = 1$.

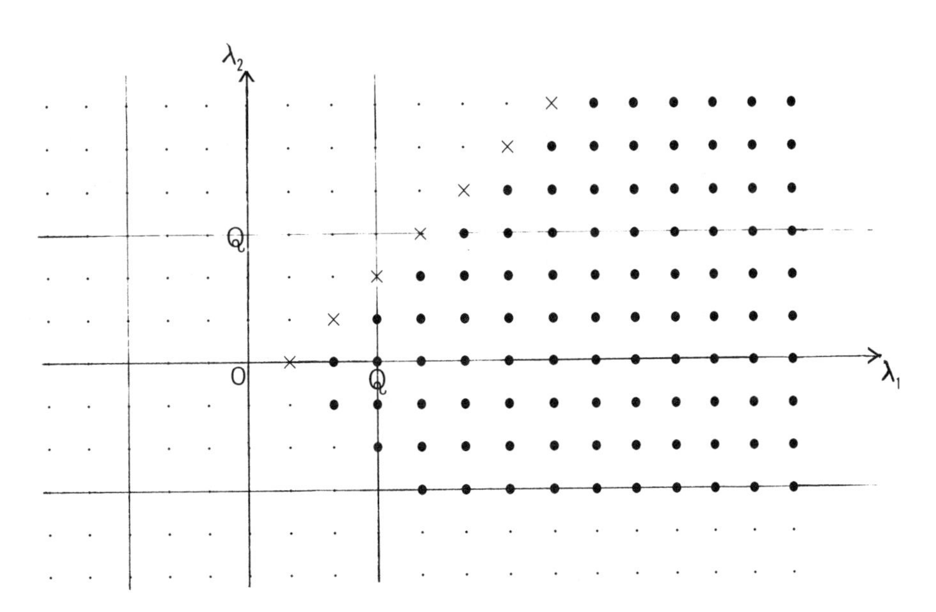

Figure 2.7.10. $(Q = p+q-2 = 3)$

$\bullet$: $\mathcal{R}_{\mathfrak{q}}^S(\mathbb{C}_\lambda)\,(\neq 0)$ is discrete series for G/H_2 given in Theorem 1 (6);

$(\lambda_1, \lambda_2) \in \mathbb{Z}^2$, $\lambda_1 > |\lambda_2|$, $\lambda_2 \ge -3$.

$\times$: $\mathcal{R}_{\mathfrak{q}}^S(\mathbb{C}_\lambda)\,(\neq 0)$ is discrete series for G/H; $\lambda_1 = \lambda_2 + 1 \in \mathbb{N}_+$.

μ_λ is $\Delta^+(\mathfrak{k})$-dominant	$\Leftrightarrow \lambda_1 > \lambda_2 \ge 2p-6$,
$\mathbb{C}_\lambda$ is weakly fair	$\Leftrightarrow \lambda_1 \ge \lambda_2 \ge 0$,
$\mathbb{C}_\lambda$ is good	$\Leftrightarrow \lambda_1 > \lambda_2 > 3$,

$\mathcal{R}_{\mathfrak{q}}^S(\mathbb{C}_\lambda)$ has singular infinitesimal character $\Leftrightarrow \lambda_1 = \pm\lambda_2$ or $\lambda_i \in \{0, \pm 1, \pm 2, \pm 3\}$.

The curious condition $\lambda_{r-2q} \geq Q+1$ in (2.2.3) occurs only if $r > 2q \geq 2$. Choose $r = 2m = 4$. Let $G = Sp(p,q) \supset H = Sp(2) \times Sp(p-2,q) \supset H_2 = Sp(p-2,q)$ with $p+q = 7$, $4 \leq p \leq 6$. In particular, $\operatorname{rank} G/H = 2$. The cross section cut out by the (λ_2, λ_4) plane makes this figure visible, revealing the essential structure.

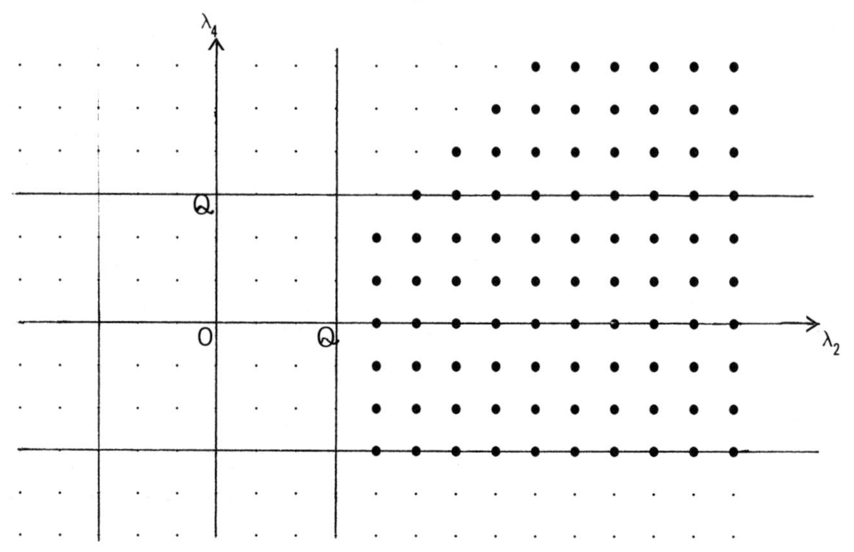

Figure 2.7.11. $(p,q) = (6,1)$, $Q = p+q-4 = 3$

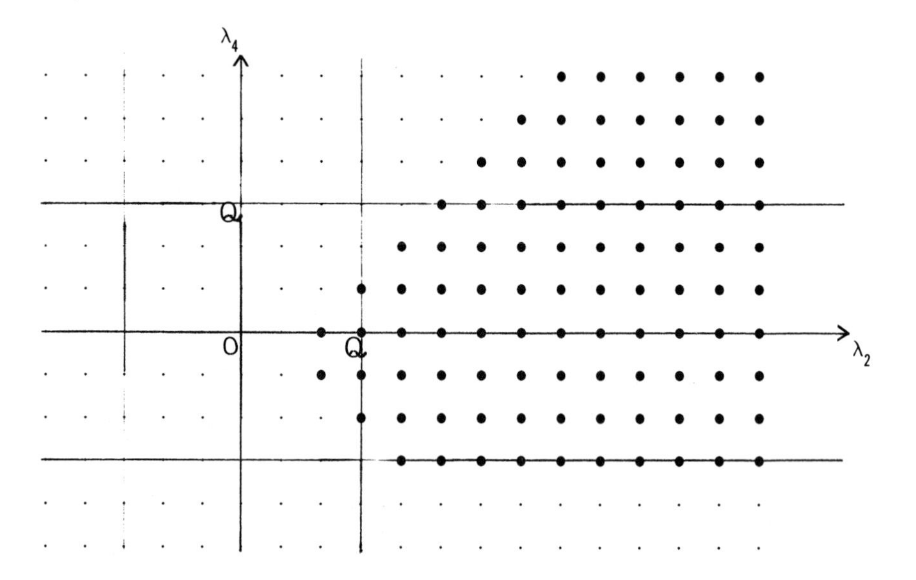

Figure 2.7.12. $(p,q) = (5,2)$ or $(4,3)$, $Q = p+q-4 = 3$

Let us explain what Figures (2.7.11) and (2.7.12) means. For each (λ_2, λ_4) corresponding to a black point • there, we have discrete series $\mathcal{R}_{\mathfrak{q}}^S(\mathbb{C}_\lambda)$ for $G/H_2 = Sp(p,q)/Sp(p-2,q)$ with parameter $\lambda = (\lambda_1, \lambda_2, \lambda_3, \lambda_4)$ whenever $(\lambda_1, \lambda_3) \in \mathbb{Z}^2$ satisfies $\lambda_1 > \lambda_2 > \lambda_3 > \lambda_4$. There exist infinitely many such (λ_1, λ_3) for each (λ_2, λ_4) corresponding to a black point •. Among them, there exists a unique choice of (λ_1, λ_3) corresponding to discrete series for a semisimple symmetric space $G/H = Sp(p,q)/Sp(2) \times Sp(p-2,q)$, namely $\lambda_1 := \lambda_2 + 1, \lambda_3 := \lambda_4 + 1$. In this case where $r = 2m = 4$, the condition $\lambda_{r-2q} \geq Q + 1$ appears iff $q = 1$, which distinguishes Figure(2.7.11) $(q = 1)$ from Figure(2.7.12) $(q \geq 2)$.

2.8. remarks

Suppose we are in the setting of Theorem 1, i.e. $G = Sp(p,q)$. We explain what is easy and what is non-trivial by existing theories. The later sections are devoted to the proof for only non-trivial parts.

Part (0) and (1) are obvious. From Fact(1.4.2)(1-a), the non trivial part in (2) is when $\mathbb{C}_\lambda$ is outside the weakly fair range, i.e. $0 > \lambda_r \geq -Q$. This case is proved in §7.1. We remark that if $-Q > \lambda_r$, then it may happen that the cohomology $\mathcal{R}_{\mathfrak{q}}^{S-j}(\mathbb{C}_\lambda) \neq 0$ for $j = 0, 1$.

As for Part (3), it is easy to understand the conditions $\lambda_i \neq \lambda_j$ because they correspond to the 'compact simple root condition' (see Lemma(3.2.1)(1)). The last condition $\lambda_{r-2q} \geq Q + 1$ is mysterious. Understanding this condition is the main subject of §4 (K-type method) and §5.

The irreducibility results (5) in the weakly fair range in Theorems 2 and 3 are derived from existing $\mathcal{D}$-module theory (Fact(6.2.4)). So we shall restrict ourselves to the case $\mathfrak{g} = \mathfrak{sp}(n, \mathbb{C})$ in §6. It is likely that all the discrete series in Part(6) are irreducible even if the fair range condition $\lambda_r > 0$ is dropped. I have checked only the case where $p = r$

by Vogan's U_α-calculus.

Finally, the condition $\lambda_r \geq -Q$ corresponds to the condition that there exists a cyclic ${H_2}^r$-fixed vector in some principal series (see (0.2)(b)), while the condition $\lambda_{r-1}+\lambda_r > 0$ corresponds to square integrability. As we said in the Introduction, the proof of Part (6) is not treated in this paper.

§3. Further notations and preliminary results

3. Jantzen-Zuckerman's translation functor

Let G be a linear connected reductive Lie group as in §1 and $H \subset G$ be a Cartan subgroup. When $\xi \in \mathfrak{h}$ is in the weight lattice for $\mathfrak{g}$ we write

$$(3.1)(a) \qquad F(\mathfrak{g}, \xi).$$

for the unique irreducible finite dimensional representation of $\mathfrak{g}$ of extremal weight ξ. Analogously, if this representation lifts to G, we denote the lift by

$$(3.1)(b) \qquad F(G, \xi).$$

This is unambiguous because G is connected. The lifting condition is that ξ is in *the weight lattice in* H *of* G, that is, the the subgroup of $\mathfrak{h}^*$ consisting of weights of finite dimensional representations of G. With a little abuse of notation, we also write $F(G_{\mathbb{C}}, \xi)$ for the holomorphic representation of $G_{\mathbb{C}}$ with extremal weight $\xi \in \mathfrak{h}^*$ in §6, and $F(L^{\sim}, \xi)$ for the metapletic representation $F(L, \xi - \rho(\mathfrak{u})) \otimes \mathbb{C}_{\rho(\mathfrak{u})}$ in the setting of §1.1.

If X is a $\mathfrak{Z}(\mathfrak{g})$-finite $\mathfrak{g}$-module and γ is a character of $\mathfrak{Z}(\mathfrak{g})$, we define

$$P_\gamma(X) := \{x \in X; (z - \gamma(z))^n x = 0 \text{ for some } n > 0, \text{ and all } z \in \mathfrak{Z}(\mathfrak{g})\}.$$

Suppose $\xi \in \mathfrak{h}^*$ is in the weight lattice, and $\gamma \in \mathfrak{h}^*$. The Jantzen-Zuckerman translation functor is defined by

$$\psi_\gamma^{\gamma+\xi}(X) := P_{\gamma+\xi}(P_\gamma(X) \otimes F(G, \xi)).$$

3.2. induction by stages

The spectral sequence of induction by stages sometimes clarifies critical points in proof. We shall frequently need the following well-known

Lemma 3.2.1 (see [28] Proposition 6.3.6). *Let $\mathfrak{q} = \mathfrak{l} + \mathfrak{u} \subset \mathfrak{p} = \mathfrak{m} + \mathfrak{n}$ be θ-stable parabolic subalgebras of $\mathfrak{g}$ with $\mathfrak{l} \subset \mathfrak{m}$. Fix a metapletic $(\mathfrak{l}, (L \cap K)^{\sim G})$-module W with respect to $\mathfrak{l} \subset \mathfrak{g}$ (see §1.1 for notation). Notice that $W \otimes \mathbb{C}_{-\rho(\mathfrak{n})}$ is a metapletic $(\mathfrak{l}, (L \cap K)^{\sim M})$-module with respect to $\mathfrak{l} \subset \mathfrak{m}$. Then there is a first quadrant spectral sequence of $(\mathfrak{g}, K)$-modules:*

$$\left(\mathcal{R}_{\mathfrak{p}}^{\mathfrak{g}}\right)^{i}\left(\mathbb{C}_{\rho(\mathfrak{n})} \otimes \left(\mathcal{R}_{\mathfrak{q}\cap\mathfrak{m}}^{\mathfrak{m}}\right)^{j}\left(W \otimes \mathbb{C}_{-\rho(\mathfrak{n})}\right)\right) \Rightarrow \left(\mathcal{R}_{\mathfrak{q}}^{\mathfrak{g}}\right)^{i+j}(W).$$

If $\mathbb{C}_{\rho(\mathfrak{n})}$ lifts to a character of M, then we can write this simply as

$$\left(\mathcal{R}_{\mathfrak{p}}^{\mathfrak{g}}\right)^{i}\left(\left(\mathcal{R}_{\mathfrak{q}\cap\mathfrak{m}}^{\mathfrak{m}}\right)^{j}(W)\right) \Rightarrow \left(\mathcal{R}_{\mathfrak{q}}^{\mathfrak{g}}\right)^{i+j}(W).$$

Finally, suppose that W has $\mathfrak{Z}(\mathfrak{l})$-infinitesimal character $\gamma \in \mathfrak{h}^*$. As a special case of the above spectral sequence, we have

1) *Assume that M/L is compact and that W is finite dimensional. If γ is not regular and integral for $\mathfrak{m}$, then $\left(\mathcal{R}_{\mathfrak{q}}^{\mathfrak{g}}\right)^{j}(W) = 0$ for all j.*

2) *If γ is in the weakly good range with respect to $\mathfrak{p} \subset \mathfrak{g}$, then*

$$\left(\mathcal{R}_{\mathfrak{p}}^{\mathfrak{g}}\right)^{s}\left(\mathbb{C}_{\rho(\mathfrak{n})} \otimes \left(\mathcal{R}_{\mathfrak{q}\cap\mathfrak{m}}^{\mathfrak{m}}\right)^{j}\left(W \otimes \mathbb{C}_{-\rho(\mathfrak{n})}\right)\right) \simeq \left(\mathcal{R}_{\mathfrak{q}}^{\mathfrak{g}}\right)^{s+j}(W).$$

where $s := \dim(\mathfrak{n} \cap \mathfrak{k})$.

3.3. definition of $\mathcal{A}(\lambda \triangleright \lambda')$

Let $\mathfrak{l}$ be the centralizer in $\mathfrak{g}$ of a semisimple abelian subspace $\mathfrak{t}$ and fix a Cartan subalgebra $\mathfrak{h} \subset \mathfrak{l}$ and a positive system $\Delta^+(\mathfrak{l}, \mathfrak{h})$. For λ, $\lambda' \in \mathfrak{h}^*$ such that $\lambda' - \lambda$ is contained in the weight lattice of $\mathfrak{g}$, we introduce a finite subset $\mathcal{A}(\Delta^+(\mathfrak{l}); \lambda \triangleright \lambda') \equiv \mathcal{A}(\lambda \triangleright \lambda')$ of $\mathfrak{h}^*$ as follows: For $\mu \in \mathfrak{h}^*$, μ is an element of $\mathcal{A}(\lambda \triangleright \lambda')$ iff

(3.3.1)(a) $\mu - \lambda$ is dominant for $\Delta^+(\mathfrak{l}, \mathfrak{h})$,

(3.3.1)(b) $\mu - \lambda$ occurs as a weight of $F(\mathfrak{g}, \lambda' - \lambda)$ for $\mathfrak{h}$,

(3.3.1)(c) $\mu + \rho_{\mathfrak{l}} \in W(\mathfrak{g}, \mathfrak{h}) \cdot (\lambda' + \rho_{\mathfrak{l}})$. (Weyl group orbit)

Inspecting the definition, one finds

$$\sigma \cdot \mathcal{A}(\Delta^+(\mathfrak{l}); \lambda \triangleright \lambda') = \mathcal{A}(\sigma \cdot \Delta^+(\mathfrak{l}); \sigma \cdot \lambda \triangleright \sigma \cdot \lambda'),$$

for $\sigma \in W(\mathfrak{l}, \mathfrak{h})$.

In later applications we will usually choose λ, $\lambda' \in \mathfrak{h}^*$ so that

$$\mathcal{A}(\lambda \triangleright \lambda') \subset \mathfrak{t}^* \tag{3.3.2}$$

as well as λ, $\lambda' \in \mathfrak{t}^*$ in the notation(1.1.1).

3.4. $\mathcal{A}(\lambda \triangleright \lambda')$ and derived functor modules

We describe a standard lemma for the connection between the translation principle and derived functor modules in terms of $\mathcal{A}(\lambda \triangleright \lambda')$. The smaller the cardinality of $\mathcal{A}(\lambda \triangleright \lambda')$ is, the less complicated the effect of translation functors on cohomologically induced representations becomes. This is the reason why we shall often use translation away from one wall to another wall avoiding regular infinitesimal characters.

Lemma(3.4.1) has two typical applications in later sections, in which we can proceed by induction on appropriate strings of λ's in proving some properties on $\mathcal{R}^i_{\mathfrak{q}}(\mathbb{C}_\lambda)$. One is to show $\mathcal{R}^{S-i}_{\mathfrak{q}}(\mathbb{C}_\lambda) = 0$ $(i \neq 0)$ where $\mathbb{C}_\lambda$ is not necessarily in the fair range (see §7). The starting point of induction here is of course in the fair (or good) range. The other is to find a sufficient condition for $\mathcal{R}^S_{\mathfrak{q}}(\mathbb{C}_\lambda) \neq 0$ (see §5), which is non-trivial even if $\mathbb{C}_\lambda$ is fair. The starting point here is sometimes outside the fair range.

Lemma 3.4.1 (cf. [28] Chapter 7 §4). *Suppose we are given two metapletic $(\mathfrak{l}, (L \cap K)^{\sim})$-modules $\mathbb{C}_\lambda$, $\mathbb{C}_{\lambda'}$ with λ, $\lambda' \in \mathfrak{t}^*$.*

1) *If $\mathcal{R}_{\mathfrak{q}}^j(\mathbb{C}_\lambda) = \mathcal{R}_{\mathfrak{q}}^j(F(L^{\sim}, \mu)) = 0$ for all $j \in \mathbb{Z}$ and all $\mu \in \mathcal{A}(\lambda \triangleright \lambda') \setminus \{\lambda'\}$, then $\mathcal{R}_{\mathfrak{q}}^j(\mathbb{C}_{\lambda'}) = 0$ for all $j \in \mathbb{Z}$.*

2) *If $\mathcal{A}(\lambda \triangleright \lambda') = \{\lambda'\}$, then*

$$\psi_{\lambda+\rho_{\mathfrak{l}}}^{\lambda'+\rho_{\mathfrak{l}}}(\mathcal{R}_{\mathfrak{q}}^j(\mathbb{C}_\lambda)) = \mathcal{R}_{\mathfrak{q}}^j(\mathbb{C}_{\lambda'}).$$

In particular, $\mathcal{R}_{\mathfrak{q}}^j(\mathbb{C}_\lambda) = 0$ implies $\mathcal{R}_{\mathfrak{q}}^j(\mathbb{C}_{\lambda'}) = 0$ for each fixed $j \in \mathbb{Z}$.

3) *Suppose that $\mathcal{A}(\lambda \triangleright \lambda')$ consists of two elements. Say, let $\mathcal{A}(\lambda \triangleright \lambda') = \{\mu, \lambda'\}$. Assume that $\mathcal{R}_{\mathfrak{q}}^j(\mathbb{C}_\lambda) = \mathcal{R}_{\mathfrak{q}}^j(F(L^{\sim}, \mu)) = 0$ for all $j \neq S$. Then,*

a) *$\mathcal{R}_{\mathfrak{q}}^j(\mathbb{C}_{\lambda'}) = 0$ for $j \neq S-1$, S.*

b) *If $\lambda' \notin \mu + \Delta(U(\mathfrak{u}), \mathfrak{h})$ and if*

$$\operatorname{Hom}_{(\mathfrak{g},K)}(\mathcal{R}_{\mathfrak{q}}^{S-1}(\mathbb{C}_{\lambda'}), \mathcal{R}_{\mathfrak{q}}^S(F(L^{\sim}, \mu))) = 0, \tag{3.4.2}$$

then $\mathcal{R}_{\mathfrak{q}}^j(\mathbb{C}_{\lambda'}) = 0$ for $j \neq S$.

b)$'$ *If $\lambda' \notin \mu + \Delta(U(\mathfrak{u}), \mathfrak{h})$ and if $\mu - \lambda$ is a unique weight of $F(G, \lambda' - \lambda)$ with the property (3.3.1)(a), then there is a long exact sequence of $(\mathfrak{g}, K)$-modules:*

$$\begin{aligned}\cdots \to \mathcal{R}_{\mathfrak{q}}^{p-1}(\mathbb{C}_{\lambda'}) \to \mathcal{R}_{\mathfrak{q}}^p(F(L^{\sim}, \mu)) \to \psi_{\lambda+\rho_{\mathfrak{l}}}^{\lambda'+\rho_{\mathfrak{l}}}(\mathcal{R}_{\mathfrak{q}}^p(\mathbb{C}_\lambda)) \\ \to \mathcal{R}_{\mathfrak{q}}^p(\mathbb{C}_{\lambda'}) \to \mathcal{R}_{\mathfrak{q}}^{p+1}(F(L^{\sim}, \mu)) \to \cdots .\end{aligned}$$

c) *If $\mu \notin \lambda' + \Delta(U(\mathfrak{u}), \mathfrak{h})$ (especially, if $\lambda' \in \mu + \Delta(U(\mathfrak{u}), \mathfrak{h})$), then $\mathcal{R}_{\mathfrak{q}}^j(\mathbb{C}_{\lambda'}) = 0$ for $j \neq S$.*

Remark 3.4.3. In Part (3) of the above lemma, we allow the multiplicity of a weight $\mu - \lambda$ in $F(G, \lambda' - \lambda)$ to be greater than one. Observe that the multiplicity of $\lambda' - \lambda$ is always one.

Proof. We can prove this Lemma by an argument similar to that in Chapter 7 §4 [28], where the $\mathfrak{Z}(\mathfrak{g})$-infinitesimal characters $\lambda+\rho_{\mathfrak{l}}$ and $\lambda'+\rho_{\mathfrak{l}}$ are assumed that one is regular and another lies in a generic point in the walls. So we review the argument there with a sketch of a necessary modification in our setting.

We first recall from Lemma(7.2.3) in [28] the following fact: Let F be a finite dimensional representation of G, and

$$\{0\} = F_0 \subset F_1 \subset \cdots \subset F_n = F \tag{3.4.4}$$

be a $(\mathfrak{q}, L\cap K)$ stable filtration with trivial induced action of $\mathfrak{u}$ on F_i/F_{i-1}. Then there is a natural spectral sequence of $(\mathfrak{g}, K)$-modules

$$\mathcal{R}^p_{\mathfrak{q}}(W\otimes F_i/F_{i-1}) \Rightarrow \mathcal{R}^p_{\mathfrak{q}}(W)\otimes F. \tag{3.4.5}$$

Assume that $W \simeq \mathbb{C}_\lambda$ $(\lambda \in \mathfrak{t}^* \subset \mathfrak{l}^*)$, $F = F(G, \lambda'-\lambda)$ and $F_i/F_{i-1} \simeq F(L,\mu_i)$ with $\mu_i \in \mathfrak{h}^*$ dominant for $\Delta^+(\mathfrak{l},\mathfrak{h})$. Applying the projection $P_{\lambda'+\rho_{\mathfrak{l}}}$ on the both sides of (3.4.5), we obtain

$$P_{\lambda'+\rho_{\mathfrak{l}}}\left(\mathcal{R}^p_{\mathfrak{q}}(F(L^\sim,\lambda+\mu_i))\right) \Rightarrow \psi^{\lambda'+\rho_{\mathfrak{l}}}_{\lambda+\rho_{\mathfrak{l}}}\left(\mathcal{R}^p_{\mathfrak{q}}(\mathbb{C}_\lambda)\right). \tag{3.4.6}$$

But $P_{\lambda'+\rho_{\mathfrak{l}}}(\mathcal{R}^p_{\mathfrak{q}}(F(L^\sim,\lambda+\mu_i))) \neq 0$ only if $\lambda+\rho_{\mathfrak{l}}+\mu_i = w\cdot(\lambda'+\rho_{\mathfrak{l}})$ for some $w \in W(\mathfrak{g},\mathfrak{h})$, thus only if $\lambda+\mu_i \in \mathcal{A}(\lambda \triangleright \lambda')$.

Recall that $\mathcal{R}^j_{\mathfrak{q}}(W) = 0$ for any $(\mathfrak{l},(L\cap K)^\sim)$ module W and any $j > S$ ([28] Corollary 6.3.21). Now a standard spectral sequence argument, combined with the following claim, completes the proof of the lemma. □

Recall that $F_{|\mathfrak{l}} \simeq \oplus_{i=1}^n F(L,\mu_i)$ as a $(\mathfrak{l}, L\cap K)$-module.

Claim 3.4.7. *With notation as above, assume ν_1, $\nu_2 \in \{\mu_1,\dots,\mu_n\}$ satisfy $\nu_2 \notin \nu_1 + \Delta(U(\mathfrak{u}))$. Then*

$$U(\mathfrak{q})F(L,\nu_1) \subsetneqq U(\mathfrak{q})(F(L,\nu_1)+F(L,\nu_2)).$$

In particular, we can choose a filtration (3.4.4) so that $i < j$ whenever $\mu_i = \nu_1$ and $\mu_j = \nu_2$.

Proof of Claim. Consider an $(\mathfrak{l}, L \cap K)$ surjective homomorphism

$$U(\mathfrak{u}) \otimes F(L, \nu_1) \to U(\mathfrak{u})F(L, \nu_1) \simeq U(\mathfrak{q})F(L, \nu_1) \subset F,$$

given by $u \otimes v \mapsto u \cdot v$ for $u \in U(\mathfrak{u})$ and $v \in F(L, \nu_1)$. The highest weight of an irreducible finite dimensional $\mathfrak{l}$ module occurring in $U(\mathfrak{u}) \otimes F(L, \nu_1)$ is of the form $\nu_1 + \delta$ with some $\delta \in \Delta(U(\mathfrak{u}))$. Therefore this remains true for the $(\mathfrak{l}, L \cap K)$-module $U(\mathfrak{q})F(L, \nu_1)$. From our assumption, we conclude that $F(L, \nu_2)$ does not occur in $U(\mathfrak{q})F(L, \nu_1)$. □

3.5. some symbols

Let $k, m \in \mathbb{C}$, and $n \in \mathbb{N}$. Write

$$[k; m, n] := (k, k + m, \ldots, k + (n-1)m) \in \mathbb{C}^n.$$

We also use the following notations:

$$\langle n \rangle := [n; -1, n] = (n, n-1, \ldots, 1),$$
$$k^n := [k; 0, n] = (k, k, \ldots, k).$$

From definition we have

$$a_1[k_1; m_1, n] + a_2[k_2; m_2, n] = [a_1 k_1 + a_2 k_2; a_1 m_1 + a_2 m_2, n],$$
$$[k; m, n_1 + n_2] = [k; m, n_1] \oplus [k + n_1 m; m, n_2],$$

for a_i, k_i, k, m_i, $m \in \mathbb{C}$ and n_i, $n \in \mathbb{N}$ $(i = 1, 2)$.

When $\lambda = (\lambda_1, \ldots, \lambda_n) \in \mathbb{C}^n$, we write ${}^t\lambda$ for $(\lambda_n, \ldots, \lambda_1)$, the transpose of λ.

The Weyl group $W(C_n) \simeq \mathfrak{S}_n \ltimes {\mathbb{Z}_2}^n$ acts by permuting and changing the signs of the coordinates of $\mathbb{C}^n$. For a fixed p $(1 \leq p \leq n)$, $1 \times W(C_{n-p})$, $W(A_n) \simeq \mathfrak{S}_n$ and so on are regarded as subgroups of $W(C_n)$ in an obvious way. These actions are sometimes restricted to an invariant subspace of $\mathbb{C}^n$.

§4. Some explicit formulas on K multiplicities

4. preliminaries

In this section, we shall give some explicit formulas about K-types occurring in $\sum_i (-1)^i \mathcal{R}_{\mathfrak{q}}^{S-i}(\mathbb{C}_\lambda)$. If the modules $\mathcal{R}_{\mathfrak{q}}^{S-i}(\mathbb{C}_\lambda)$ vanish except in a single degree $i = 0$, this formula determines immediately whether $\mathcal{R}_{\mathfrak{q}}^{S}(\mathbb{C}_\lambda) = 0$ or not. A condition guaranteeing the vanishing of $\mathcal{R}_{\mathfrak{q}}^{S-i}(\mathbb{C}_\lambda)$ $(i \neq 0)$ outside the fair range (cf. Fact(1.4.2)) will be given in §7 when $\mathcal{R}_{\mathfrak{q}}^{S}(\mathbb{C}_\lambda)$ corresponds to discrete series for $U(p,q;\mathbb{F})/U(p-m,q;\mathbb{F})$ (see §2). The setting of this section will be slightly more general in the range of $\mathbb{C}_\lambda$ than the results in §2.

Suppose we are in the setting of §1.1. Let $\mathfrak{h} = \mathfrak{t}^c + \mathfrak{a}^c$ be a fundamental Cartan subalgebra of $\mathfrak{g}$ with $\mathfrak{h} \subset \mathfrak{l}$. As usual we write half the sum of roots as $\rho_c \equiv \rho(\mathfrak{k}, \mathfrak{t}^c)$, $\rho(\mathfrak{u}) \equiv \rho(\mathfrak{u}, \mathfrak{h})$ and $\rho(\mathfrak{u} \cap \mathfrak{k}) \equiv \rho(\mathfrak{u} \cap \mathfrak{k}, \mathfrak{t}^c)$, respectively. Let $\mu_\lambda := (\lambda + \rho(\mathfrak{u}) - 2\rho(\mathfrak{u} \cap \mathfrak{k}))_{|\mathfrak{t}^c} \in \mathfrak{t}^{c*}$. We remark that our definition of $\mathcal{R}_{\mathfrak{q}}^{S-i}(\mathbb{C}_\lambda)$ here follows [32] Definition 6.20 and differs from [28] Definition 6.3.1 by a $\rho(\mathfrak{u})$-shift (see §1.3). Recall that $S = \dim(\mathfrak{u} \cap \mathfrak{k})$. Then the Blattner formula due to Hecht-Schmid and generalized by Vogan ([28] Theorem 6.3.12) gives

$$(4.1) \quad \sum_i (-1)^i \dim \operatorname{Hom}_K \left(\pi, \mathcal{R}_{\mathfrak{q}}^{S-i}(\mathbb{C}_\lambda)\right)$$
$$= \sum_j (-1)^j \dim \operatorname{Hom}_{L \cap K} \left(H^j(\mathfrak{u} \cap \mathfrak{k}, \pi), S(\mathfrak{u} \cap \mathfrak{p}) \otimes \mathbb{C}_{\mu_\lambda}\right)$$

for $\pi \in \widehat{K}$, where $S(V)$ denotes a symmetric tensor algebra of a vector space V. When μ_λ is $\Delta^+(\mathfrak{k})$ dominant, it is easy to see that the multiplicity of the particular K-type

$\pi = F(K,\mu_\lambda)$ (notation (3.1.1)) in $\sum_i(-1)^i\mathcal{R}_{\mathfrak{q}}^{S-i}(\mathbb{C}_\lambda)$ equals one, from formula (4.1.1), whence $\sum_i(-1)^i\mathcal{R}_{\mathfrak{q}}^{S-i}(\mathbb{C}_\lambda) \neq 0$. However, for actual calculation with more general λ or π, too many cancellations occur, so that it is hard to tell even whether the formula(4.1.1) vanishes or not. T.Oshima showed me an unpublished note (Lemma(4.2.6)) on how to calculate the Blattner formula for some particular small K type (which frequently coincides with one of the minimal K-types in Vogan's sense, cf. [28]) occurring in the discrete series for a semisimple symmetric space $G/H = SO(p,q)/SO(m)\times SO(p-m,q)$ even if μ_λ is not $\Delta^+(\mathfrak{k})$ dominant. Inspired by his technique, we are able to control *all* K-types in terms of some alternating polynomial functions (see (4.2.2)). To do this we also need another idea about 'collecting K-types'. Communications with H.Schlichtkrull clarified this idea in the following general setting. Now let us explain it roughly.

Retain notations in §1.1-3. (So $\mathfrak{t}_0$ is a fixed abelian subalgebra of $\mathfrak{k}_0$ and $\mathfrak{h}_0$ is a fundamental Cartan subalgebra of $\mathfrak{l}_0$ (and also of $\mathfrak{g}_0$) such that $\mathfrak{t} \subset \mathfrak{t}^c = \mathfrak{k}\cap\mathfrak{h} \subset \mathfrak{h} \subset \mathfrak{l}$.) We may assume that a θ-stable parabolic subalgebra $\mathfrak{q} = \mathfrak{l}+\mathfrak{u}$ is defined by

$$\Delta(\mathfrak{u},\mathfrak{h}) = \{\alpha \in \Delta(\mathfrak{g},\mathfrak{h})\,;\ \langle\alpha,\,\nu\rangle \geq 1\},$$

$$\Delta(\mathfrak{l},\mathfrak{h}) = \{\alpha \in \Delta(\mathfrak{g},\mathfrak{h})\,;\ \langle\alpha,\,\nu\rangle = 0\},$$

with a normalized generic element $\nu \in \sqrt{-1}\mathfrak{t}_0^*$.

We choose a positive system of $\Delta(\mathfrak{k},\mathfrak{t}^c)$ so that $\Delta(\mathfrak{u}\cap\mathfrak{k}) \subset \Delta^+(\mathfrak{k})$. Set $\Theta_\lambda := \sum_i(-1)^i\mathcal{R}_{\mathfrak{q}}^{S-i}(\mathbb{C}_\lambda)$ and $\pi = F(K,\mu)$. The information of *each* K-type occurring in Θ_λ (e.g. $[\Theta_{\lambda|K} : \pi] = \sum_j(-1)^j \dim \operatorname{Hom}_K(\pi, \mathcal{R}_{\mathfrak{q}}^{S-j}(\mathbb{C}_\lambda))$ is so detailed that we can hardly expect that it is a nice function of $\mu \in (\mathfrak{t}^c)^*$ and $\lambda \in \mathfrak{t}^*$. In fact, such a function may behave somewhat irregularly when λ and μ are small enough. So our idea is to investigate more course information about K-types:

Set

$$\widehat{K}(\Theta_\lambda) := \{\pi \in \widehat{K}\,;\, [\Theta_{\lambda|K} : \pi] \neq 0\},$$

$$\widehat{K}(\delta) := \{\pi \in \widehat{K}\,;\, \text{the highest weight } \mu \text{ of } \pi \text{ satisfies } \mu_{|\mathfrak{t}} = \delta\},$$

for $\lambda \in \mathfrak{t}^*$ and $\delta \in \mathfrak{t}^*$. If λ does not satisfy the integrality conditions so that $\mathbb{C}_\lambda$ lifts to a metapletic $(\mathfrak{l}, (L \cap K)^\sim)$-character, we set $\widehat{K}(\Theta_\lambda) := \{\emptyset\}$. Similarly $\widehat{K}(\delta) := \{\emptyset\}$ if δ does not lie in the obvious lattice in $\sqrt{-1}\mathfrak{t}_0^*$. Clearly,

$$\Theta_{\lambda|K} \equiv 0 \iff \widehat{K}(\Theta_\lambda) = \emptyset,$$

$$\widehat{K} = \coprod_{\delta \in \mathfrak{t}^*} \widehat{K}(\delta).$$

Writing $\mathcal{P}(\widehat{K})$ for the totality of subsets in $\widehat{K}$, we have a map

$$\mathfrak{t}^* \times \mathfrak{t}^* \ni (\lambda, \delta) \mapsto \widehat{K}(\Theta_\lambda) \cap \widehat{K}(\delta) \in \mathcal{P}(\widehat{K}).$$

Take a positive-valued function $d\colon \widehat{K} \to \mathbb{N}_+$ (e.g. $d(\pi) \equiv 1$, or $d(\pi) := \dim \pi$). Now we define a map $M\colon \mathfrak{t}^* \times \mathfrak{t}^* \to \mathbb{Z}$ by

$$M(\lambda, \delta) := \sum_{\pi \in \widehat{K}(\delta)} d(\pi)\,[\Theta_{\lambda|K} : \pi] = \sum_{\pi \in \widehat{K}(\Theta_\lambda) \cap \widehat{K}(\delta)} d(\pi)\,[\Theta_{\lambda|K} : \pi]. \tag{4.1.2}$$

If $\mathcal{R}_{\mathfrak{q}}^i(\mathbb{C}_\lambda)$ vanishes except in a single degree, then it follows from definition that

$$M(\lambda, \delta) = 0 \text{ for all } \delta \in \mathfrak{t}^* \iff \mathcal{R}_{\mathfrak{q}}^i(\mathbb{C}_\lambda) = 0 \text{ for all } i.$$

What we would expect now is that, with a suitable choice of $d\colon \widehat{K} \to \mathbb{N}_+$, *this* $M(\lambda, \delta)$ *is a restriction of a nice polynomial function of* δ *and* λ *so that we can tell explicitly whether or not* $M(\lambda, \delta) = 0$ *for all* $\delta \in \mathfrak{t}^*$. One of the simplest examples is a ladder representation which has K-highest weights lying along a single line ([7]) and in this case $\dim \mathfrak{t} = 1$. We will give a beautiful formula of $M(\lambda, \delta)$ with a special choice of d in some settings (see (4.1.8) for precise definition). The formula is expressed as a determinant of a certain matrix whose entries are polynomials of λ and δ, and the matrix reduces to a scalar if $\dim \mathfrak{t} = 1$. First of all, let us check that Definition(4.1.2) always makes sense. This is guaranteed by the following

Proposition 4.1.3. *Retain notations as above. Then,*

$$\sharp\left(\widehat{K}(\Theta_\lambda) \cap \widehat{K}(\delta)\right) < \infty.$$

We note that $\widehat{K}(\delta)$ is possibly an infinite set, as is $\widehat{K}(\Theta_\lambda)$. The proof of Proposition(4.1.3) is based on a uniform estimate over $\widehat{K}(\delta)$ as used in calculating the Blattner formula. That is,

Lemma 4.1.4. *With notations as above, fix δ, $b \in \sqrt{-1}\mathfrak{t}_0^*$. Recall that ν is a fixed element of $\sqrt{-1}\mathfrak{t}_0^*$, defining a θ-stable parabolic subalgebra $\mathfrak{q} = \mathfrak{l} + \mathfrak{u}$. If $\pi \in \widehat{K}(\delta)$ satisfies* $\operatorname{Hom}_T(H^j(\mathfrak{u} \cap \mathfrak{k}, \pi), S^m(\mathfrak{u} \cap \mathfrak{p}) \otimes \mathbb{C}_b) \neq 0$, *then*

$$m \leq \langle \delta - b, \nu \rangle - j.$$

Postponing the proof of Lemma(4.1.4) for a while, we first prove Proposition(4.1.3).

Proof of Proposition(4.1.3). Let $\pi_\mu \in \widehat{K}(\delta)$. From (4.1.1) and Lemma(4.1.4),

$$\begin{aligned} \|[\Theta_\lambda : \pi_\mu]\| &\leq \sum_j \dim \operatorname{Hom}_{L \cap K}(H^j(\mathfrak{u} \cap \mathfrak{k}, \pi), \bigoplus_{m=0}^{\infty} S^m(\mathfrak{u} \cap \mathfrak{p}) \otimes \mathbb{C}_{\mu_\lambda}) \\ &\leq \sum_j \dim \operatorname{Hom}_{L \cap K}(H^j(\mathfrak{u} \cap \mathfrak{k}, \pi), \bigoplus_{m=0}^{\langle \delta - \mu_\lambda, \nu \rangle} S^m(\mathfrak{u} \cap \mathfrak{p}) \otimes \mathbb{C}_{\mu_\lambda}). \end{aligned}$$

Therefore, we have

$$\widehat{K}(\Theta_\lambda) \cap \widehat{K}(\delta) \subset \bigcup_{j,\tau} \{\pi \in \widehat{K}\,;\, \operatorname{Hom}_{L \cap K}(H^j(\mathfrak{u} \cap \mathfrak{k}, \pi), \tau) \neq 0\},$$

where τ runs over the finite subset of $\widehat{L \cap K}$ occurring in $\bigoplus_{m=0}^{\langle \delta - \mu_\lambda, \nu \rangle} S^m(\mathfrak{u} \cap \mathfrak{p}) \otimes \mathbb{C}_{\mu_\lambda}$. From Kostant's Borel-Weil-Bott theorem ([15], see also [28] Corollary 3.2.16), the highest weight of an irreducible representation of $L \cap K$ occurring in $H^j(\mathfrak{u} \cap \mathfrak{k}, \pi)$ is of the form

$$w \cdot (\mu + \rho_c) - \rho_c \in (\mathfrak{t}^c)^*,$$

with $w \in W_K^{\mathfrak{l} \cap \mathfrak{k}}$ and $l(w) = j$. Here

$$W_K^{\mathfrak{l} \cap \mathfrak{k}} := \{w \in W(\mathfrak{k}, \mathfrak{t}^c) ; \Delta^+(w) \subset \Delta(\mathfrak{u} \cap \mathfrak{k})\},$$
$$\Delta^+(w) := \Delta^+(\mathfrak{k}) \cap w \cdot \Delta^-(\mathfrak{k}),$$
$$l(w) = \sharp \Delta^+(w).$$

Write $\eta \in (\mathfrak{t}^c)^*$ for the highest weight of $\tau \in \widehat{L \cap K}$ (recall that $L \cap K$ is of maximal rank in K). Then the equation $w \cdot (\mu + \rho_c) - \rho_c = \eta$ determines μ uniquely for each w and τ. Thus, for any $\tau \in \widehat{L \cap K}$, we have

$$\sum_j \sharp\{\pi \in \widehat{K} ; \operatorname{Hom}_{L \cap K}(H^j(\mathfrak{u} \cap \mathfrak{k}, \pi), \tau) \neq 0\} \leq \sharp W_K^{\mathfrak{l} \cap \mathfrak{k}} < \infty,$$

whence $\sharp \left(\widehat{K}(\Theta_\lambda) \cap \widehat{K}(\delta) \right) < \infty$. □

Remark 4.1.5. Define a subset of $\widehat{K}$ by

$$\widehat{K}_{(\nu)}(\delta) := \{\pi \in \widehat{K} ; \text{the highest weight } \mu \text{ of } \pi \text{ satisfies } \langle \mu, \nu \rangle = \langle \delta, \nu \rangle\}.$$

Then $\widehat{K}_{(\nu)}(\delta) \supset \widehat{K}(\delta)$. The above proof actually has shown the following statement:

$$\sharp \left(\widehat{K}(\Theta_\lambda) \cap \widehat{K}_{(\nu)}(\delta) \right) < \infty.$$

This might be used to give some variations of the definition of $M(\lambda, \delta)$ in (4.1.2).

Proof of Lemma(4.1.4). As in the above proof, any $\mathfrak{t}$-weight occurring in $H^j(\mathfrak{u} \cap \mathfrak{k}, \pi)$ is of the form

$$(w \cdot (\mu + \rho_c) - \rho_c)_{|\mathfrak{t}}$$

with $w \in W_K^{\mathfrak{l}\cap\mathfrak{k}}$ and $l(w) = j$. Since $\mu \in (\mathfrak{t}^c)^*$ is $\Delta^+(\mathfrak{k})$ dominant, there are non-negative real numbers $\{n_\alpha\,;\, \alpha \in \Delta^+(\mathfrak{k})\}$ such that $\mu - w\cdot\mu = \sum_{\alpha\in\Delta^+(\mathfrak{k})} n_\alpha\alpha$. We also have $\rho_c - w\cdot\rho_c = \sum_{\alpha\in\Delta^+(w)}\alpha$ from definition of $\Delta^+(w)$. Thus, taking an inner product with $\nu \in \sqrt{-1}\mathfrak{t}_0^* \subset (\mathfrak{t}^c)^*$, we have

$$\begin{aligned}\langle\delta,\nu\rangle = \langle\mu,\nu\rangle &= \langle(\mu - w\cdot\mu) + (\rho_c - w\cdot\rho_c) + w\cdot(\mu+\rho_c) - \rho_c, \nu\rangle\\ &= \langle \sum_{\alpha\in\Delta^+(\mathfrak{k})} n_\alpha\alpha + \sum_{\alpha\in\Delta^+(w)}\alpha + w\cdot(\mu+\rho_c) - \rho_c, \nu\rangle\\ &\geq j + \langle w\cdot(\mu+\rho_c) - \rho_c, \nu\rangle. \end{aligned} \tag{4.1.6}$$

In the last inequality, we have used $\Delta^+(\mathfrak{k}) \subset \Delta(\mathfrak{l}) \cup \Delta(\mathfrak{u})$ and $\Delta^+(w) \subset \Delta(\mathfrak{u}\cap\mathfrak{k})$.

On the other hand, any $\mathfrak{t}$-weight of $S^m(\mathfrak{u}\cap\mathfrak{p})\otimes\mathbb{C}_b$ is of the form

$$\sum_{\alpha\in\Delta(\mathfrak{u}\cap\mathfrak{p})} m_\alpha \alpha_{|\mathfrak{t}} + b$$

with some $m_\alpha \in \mathbb{N}$ satisfying $\sum m_\alpha = m$. In particular,

$$\langle \sum_{\alpha\in\Delta(\mathfrak{u}\cap\mathfrak{p})} m_\alpha\alpha_{|\mathfrak{t}} + b, \nu\rangle \geq m + \langle b,\nu\rangle. \tag{4.1.7}$$

From (4.1.6) and (4.1.7), $\mathrm{Hom}_T(H^j(\mathfrak{u}\cap\mathfrak{k},\pi), S^m(\mathfrak{u}\cap\mathfrak{p})\otimes\mathbb{C}_b) \neq 0$ implies $m + \langle b,\nu\rangle \leq \langle\delta,\nu\rangle - j$. $\square$

From now on, let us restrict ourselves to the following settings: Suppose that K has a direct decomposition $K = K_1 \times K_2$ and $\mathfrak{t} \subset \mathfrak{k}_1$. We write $\mathfrak{t}^c = \mathfrak{t}_1^c + \mathfrak{t}_2^c$ according to $\mathfrak{k} = \mathfrak{k}_1 + \mathfrak{k}_2$. Clearly, we have $\mathfrak{t} \subset \mathfrak{t}_1^c$. Let $\pi = \pi_1 \boxtimes \pi_2 \in \widehat{K} \simeq \widehat{K_1}\times\widehat{K_2}$. Putting $d(\pi) := \dim\pi_2$ in (4.1.2), we introduce a multiplicity function

$$M(\mathfrak{q},\lambda,\delta) := \sum_{\pi\in\widehat{K}(\delta)} \dim(\pi_2)\,[\Theta_{\lambda|K} : \pi], \tag{4.1.8}$$

for λ, $\delta \in \mathfrak{t}^*(\subset (\mathfrak{t}_1^c)^*)$.

The rest of this section will be devoted to showing that this $M(\mathfrak{q}, \lambda, \delta)$ behaves as a nice polynomial of δ under some hypotheses on λ (see §4.3-5 for results) when $G = U(p, q; \mathbb{F})$. We finish this subsection with another formula for $M(\mathfrak{q}, \lambda, \delta)$.

Set

$$m(\mathfrak{q}, \lambda, \pi_1) := \sum_{i=0}^{S} (-1)^i \dim \mathrm{Hom}_{K_1}(\pi_1, \mathcal{R}_{\mathfrak{q}}^{S-i}(\mathbb{C}_\lambda)). \tag{4.1.9}$$

Since $H^j(\mathfrak{u} \cap \mathfrak{k}, \pi) \simeq H^j(\mathfrak{u} \cap \mathfrak{k}_1, \pi_1) \boxtimes \pi_2$ as $L \cap K \simeq (L \cap K_1) \times K_2$ module, we have

$$m(\mathfrak{q}, \lambda, \pi_1) = \sum_j \sum_{m=0}^{\infty} (-1)^j \dim \mathrm{Hom}_{L \cap K_1}(H^j(\mathfrak{u} \cap \mathfrak{k}_1, \pi_1), S^m(\mathfrak{u} \cap \mathfrak{p}) \otimes \mathbb{C}_{\mu_\lambda}). \tag{4.1.10}$$

If $\delta^{(1)} \in (\mathfrak{t}_1^c)^*$ is the highest weight of $\pi_1 \in \widehat{K_1}$, we sometimes write $m(\mathfrak{q}, \lambda, \delta^{(1)})$ for $m(\mathfrak{q}, \lambda, \pi_1)$. Now we have

$$M(\mathfrak{q}, \lambda, \delta) = \sum_{\pi_1} m(\mathfrak{q}, \lambda, \pi_1), \tag{4.1.11}$$

where the sum is taken over the set:

$$\{\pi_1 \in \widehat{K_1}\,;\ \text{the restriction of the highest weight of } \pi_1 \text{ to } \mathfrak{t} \text{ is } \delta\}.$$

4.2. some alternating polynomials

Fix positive integers n and l. Define a polynomial of t by

$$a(t, n) \overset{\text{def}}{=} \frac{\Gamma(n+t)}{\Gamma(n)\Gamma(1+t)} = \frac{(t+n-1)\cdots(t+1)}{(n-1)!}. \tag{4.2.1}$$

and polynomials of $x = (x_1, \ldots, x_l)$ and $y = (y_1, \ldots, y_l)$ by

(4.2.2)(a) $$F(n, l; x, y) \overset{\text{def}}{=} \det\left(a(x_i + y_j; n)_{1 \le i,j \le l}\right),$$

(4.2.2)(b) $$d(n, l; x) \overset{\text{def}}{=} \det\left(a(x_i - j; n)_{1 \le i,j \le l}\right).$$

By definition we have

$$d(n,l;x) = F(n,l;x,(-1,-2,\dots,-l)). \tag{4.2.3}$$

If $t \in \mathbb{Z}$ and $t \geq -n+1$, then

$$a(t,n) = \dim S^t(\mathbb{C}^n), \tag{4.2.4}$$

where $\dim S^t(\mathbb{C}^n) := 0$ for $t < 0$. The point of (4.2.4) is that it covers small negative integers because $a(t,n) = 0$ for $-n+1 \leq t \leq -1$. This will be useful for later applications (see §4.3-5).

Lemma 4.2.5. *As a polynomial of x and y,*

$$F(n,l;x,y) \equiv 0 \quad \textit{if and only if} \quad n < l.$$

The following lemma is due to T.Oshima and J.Sekiguchi. We include here its proof, which is also due to them, for the sake of completeness. The author is grateful to them.

Lemma 4.2.6(Oshima-Sekiguchi)**.**

$$d(n,l;x) = \begin{cases} \dfrac{\prod_{i=1}^{l}\prod_{k=1}^{n-l}(k+x_i-1)}{\prod_{j=1}^{l}(n-l+j-1)!} \times \prod_{i>j}(x_i - x_j) & \text{if } n \geq l, \\ 0 : & \text{if } n < l. \end{cases}$$

In particular, for fixed n, $l \in \mathbb{N}_+$ and $x \in \mathbb{C}^l$,

$$d(n,l;x) \neq 0 \quad \textit{iff} \quad \begin{cases} n \geq l, \\ x_i \notin \{0,-1,\dots,1-n+l\} \ \text{for any } i, \\ x_i \neq x_j \quad \text{for any } i \neq j. \end{cases}$$

Proof of Lemma 4.2.5. Since $F(n, l; x, y)$ is an alternating polynomial function of (say) x, it is divisible by the simplest alternating polynomial $\prod_{i>j}(x_i - x_j)$. On the other hand, $F(n, l; x, y)$ is of degree $\leq n-1$ with respect to x_1 (expand the determinant along the first column), while $\prod_{i>j}(x_i - x_j)$ is of degree $l-1$. Hence the 'if' part of the lemma. The 'only if' part is guaranteed by the special values given in Lemma(4.2.6). □

Proof of Lemma(4.2.6). From definition, we have

(4.2.7)(a)
$$a(t, n) - a(t-1, n) = \begin{cases} a(t, n-1) & \text{if } n \geq 2, \\ 0 & \text{if } n = 1. \end{cases}$$

(4.2.7)(b)
$$a(t, n+1) = \frac{n+t}{n} a(t, n)$$

We put $a(t, n) \equiv 0$ if $n \in -\mathbb{N}$, then (4.2.7)(a) can be rewritten for $n \in \mathbb{Z}$:

(4.2.8)
$$a(x_i - j, n) = a(x_i - (j+1), n) + a(x_i - j, n-1).$$

If we apply (4.2.8) to all j $(1 \leq j \leq l-1)$ and iterate on (4.2.8) then we conclude

$$d(n, l; x) = \begin{cases} \det\left(a(x_i - j, n - l + j)_{1 \leq i, j \leq l}\right) & \text{if } n \geq l, \\ 0 & \text{if } n < l. \end{cases}$$

From now on, assume that $n \geq l$. Iteration on (4.2.7)(b) yields

$$a(x_i - j, n - l + j) = \left(\prod_{k=1}^{n-l} \frac{k + x_i - 1}{k + j - 1}\right) a(x_i - j, j)$$
$$= \left(\prod_{k=1}^{n-l} (k + x_i - 1)\right) \frac{(x_i - 1) \cdots (x_i - j + 1)}{(n - l + j - 1)!}.$$

Therefore

$$d(n,l;x) = \frac{\prod_{i=1}^{l}\prod_{k=1}^{n-l}(k+x_i-1)}{\prod_{j=1}^{l}(n-l+j-1)!} \det\left((x_i-1)\cdots(x_i-j+1)_{1\le i,j\le l}\right)$$

$$= \frac{\prod_{i=1}^{l}\prod_{k=1}^{n-l}(k+x_i-1)}{\prod_{j=1}^{l}(n-l+j-1)!} \det\left((x_i^{j-1})_{1\le i,j\le l}\right)$$

Vandermonde's determinant equals the simplest alternating polynomial, which leads to the desired formula. □

Finally we prove the following lemma for use in §4.8.

Lemma 4.2.8. *Let*

$$d'(n,l;x) := \det\left(a(x_i+j;n)_{1\le i,j\le l}\right). \quad \text{(cf. (4.2.2)(b))} \tag{4.2.9}$$

Then we have

$$d'(n,l;x) = (-1)^{l(n-1)} d(n,l;-x-[n;0,l]).$$

(Recall notation(3.5).) In particular,

$$d'(n,l;x)\neq 0 \quad \text{iff} \quad \begin{cases} n\ge l, \\ x_i\notin\{-l-1,-l-2,\dots,-n\} & \text{for any } i, \\ x_i\neq x_j \quad \text{for any } i\neq j. \end{cases}$$

Proof. This is a direct consequence of combinatorial reciprocity:

$$a(t,n) = (-1)^{n-1}a(-t-n,n). \tag{4.2.10}$$

The last statement is a translation of Lemma(4.2.6) via (4.2.10). □

4.3. result in quaternionic case

Retain notations in §4.1-2 and §2.1. We have $\mathfrak{t}^c = \mathfrak{h} \simeq \mathbb{C}^{p+q}$ in this case. Then,

$$\begin{aligned}
\rho_c &= [p;-1,p] \oplus [q;-1,q] &&= (p,p-1,\dots,1,q,q-1,\dots,1),\\
\rho(\mathfrak{u}) &= [p+q;-1,r] \oplus [0;0,p+q-r] &&= (p+q,\dots,p+q-r+1,0,\dots,0),\\
\rho(\mathfrak{u}\cap\mathfrak{k}) &= [p;-1,r] \oplus [0;0,p+q-r] &&= (p,p-1,\dots,p-r+1,0,\dots,0).
\end{aligned}$$

Thus we have,

$$\begin{aligned}
\rho(\mathfrak{u}) - 2\rho(\mathfrak{u}\cap\mathfrak{k}) &= [-p+q;1,r] \oplus [0;0,p+q-r]\\
&= (-p+q,-p+q+1,\dots,-p+q+r-1,0,\dots,0).
\end{aligned}$$

For $\lambda := (\lambda_1,\dots,\lambda_r,0,\dots,0) \in \mathfrak{h}^*$, we set

$$\mu_\lambda := \lambda + \rho(\mathfrak{u}) - 2\rho(\mathfrak{u}\cap\mathfrak{k}) = (b_1,\dots,b_r,0,\dots,0),$$

where

$$b_i := \lambda_i - p + q + i - 1 \qquad (1 \le i \le r). \tag{4.3.1}$$

Proposition 4.3.2. *Retain notations as above. Let* $\lambda_i \in \mathbb{Z}\,(1 \le i \le r)$, $\delta \in \mathbb{Z}^r$. *Recall that we assign* $b = (b_1,\dots,b_r)$ *to* λ *by* (4.3.1) *and that* $Q = p+q-r$. *Assume that*

(4.3.3)(a) $$Q \ge \lambda_1 > \lambda_2 > \cdots > \lambda_r \ge -Q,$$

(4.3.3)(b) $$\delta_1 \ge \cdots \ge \delta_r \ge 0.$$

Define $y \equiv y(\lambda) = (y_1,\dots,y_r) \in \mathbb{Z}^r$, $z \equiv z(\delta) = (z_1,\dots,z_r) \in \mathbb{Z}^r$ *by* $y_i := i - b_i = -\lambda_i + p - q + 1$ *and* $z_j := \delta_j - j$. *Set*

(4.3.4)(a) $$k := \max[\{i; 1 \le i \le r,\ b_i \ge 0\} \cup \{0\}].$$

(4.3.4)(b) $$\tilde{\mu_\lambda} := (b_1,\dots,b_k,0,\dots,0) \in \mathbb{Z}^r \subset \mathfrak{t}^*$$

Then the following holds.

1) $M(\mathfrak{q},\lambda,\delta) = F(2q,r;y(\lambda),z(\delta))$.

2) *Assume moreover that* $\sum_{i=1}^{r}\delta_i = \sum_{i=1}^{k} b_i$. *Then*

   a) *If* $\delta \neq \tilde{\mu_\lambda}$, *then* $M(\mathfrak{q},\lambda,\delta) = 0$.

   b) *If* $\delta = \tilde{\mu_\lambda}$, *then* $M(\mathfrak{q},\lambda,\delta) = d(2q, r-k; y_{k+1}(\lambda),\ldots,y_r(\lambda))$.

3) $M(\mathfrak{q},\lambda,\delta) = 0$ *if* $\sum_{i=1}^{r}\delta_i < \sum_{i=1}^{k} b_i$.

Remark 4.3.5. In the above Proposition the assumption $Q \geq \lambda_1$ in (4.3.3)(a) can be relaxed to $Q \geq \lambda_{k+1}$ in (2) and can be dropped in (3).

The proof of this Proposition together with Remark(4.3.5) will be given in §4.7 after some preparations in §4.6. We observe that $r - 2q > 0$ iff $r - 2q > k$ under the assumption (4.3.3)(a). Indeed, $b_{r-2q} = \lambda_{r-2q} - Q - 1$ when $r > 2q$.

Since $y_i(\lambda) = i - b_i \geq 1 (k+1 \leq i \leq r)$, combining Proposition(4.3.2) with Lemma(4.2.5-6), we have

Corollary 4.3.6. *Retain the same notation as in Proposition*(4.3.2). *If* $\lambda \in \mathbb{Z}^r \oplus 0^{p+q-r}$ *satisfies* (4.3.3)(a) *and if* $\mathcal{R}_{\mathfrak{q}}^{S-i}(\mathbb{C}_\lambda) = 0$ *for* $i \neq 0$, *then the following two conditions are equivalent:*

   a) $\mathcal{R}_{\mathfrak{q}}^{S}(\mathbb{C}_\lambda) = 0$.

   b) $2q < r$.

If we use induction by stages (see an argument in Remark(5.1.5) in the next section), we have immediately

Corollary 4.3.6′. *Retain the same notation as in Proposition*(4.3.2). *If* $\lambda \in \mathbb{Z}^r \oplus 0^{p+q-r}$ *satisfies*

$$\lambda_1 > \lambda_2 > \cdots > \lambda_r \geq -Q,$$

and if $\mathcal{R}_{\mathfrak{q}}^{S-i}(\mathbb{C}_\lambda) = 0$ *for* $i \neq 0$, *then the following two conditions are equivalent:*

a) $\mathcal{R}_{\mathfrak{q}}^{S}(\mathbb{C}_\lambda) = 0$.

b) $2q < r$ *and* $\lambda_{r-2q} \leq Q$.

Remark 4.3.8. μ_λ is a K-dominant weight only if $k = r$ (see (4.3.4)(a) for definition). (It is always the case when λ is 'sufficiently' regular.) This is the reason why we use $\tilde{\mu_\lambda}(\in \mathfrak{t}^*)$ instead of μ_λ.

4.4. result in complex case

Retain notations in §4.1-2 and §2.1. Note that $\mathfrak{t}^c = \mathfrak{h}$ in this case.

We abbreviate $[k; m, n]$ as $[k; m]$ when this vector is understood to be contained in $\mathbb{C}^n$. According to the direct decomposition:

$$\mathfrak{h}^* \simeq \mathbb{C}^{p+q} = \mathbb{C}^r \oplus \mathbb{C}^s \oplus \mathbb{C}^{p-r-s} \oplus \mathbb{C}^q$$

defined by basis $\{f_i\,;\, 1 \leq i \leq p+q\}$, we have

$$\rho_c = [\frac{p-1}{2}; -1] \oplus [\frac{-p+2s-1}{2}; -1] \oplus [\frac{p-2r-1}{2}; -1] \oplus [\frac{q-1}{2}; -1],$$

$$\rho(\mathfrak{u}) = [\frac{p+q-1}{2}; -1] \oplus [\frac{-p-q+2s-1}{2}; -1] \oplus [\frac{-r+s}{2}; 0] \oplus [\frac{-r+s}{2}; 0],$$

$$\rho(\mathfrak{u} \cap \mathfrak{k}) = [\frac{p-1}{2}; -1] \oplus [\frac{-p+2s-1}{2}; -1] \oplus [\frac{-r+s}{2}; 0] \oplus [0; 0].$$

Thus, $\rho(\mathfrak{u}) - 2\rho(\mathfrak{u} \cap \mathfrak{k})$

$$= [\frac{-p+q+1}{2}; 1] \oplus [\frac{p-q-2s+1}{2}; 1] \oplus [\frac{r-s}{2}; 0] \oplus [\frac{-r+s}{2}; 0].$$

For $\lambda := (\lambda_1, \ldots, \lambda_{r+s}, 0, \ldots, 0) + [\frac{-r+s}{2}; 0, p+q]$, we set

$$\mu_\lambda := \lambda + \rho(\mathfrak{u}) - 2\rho(\mathfrak{u} \cap \mathfrak{k}) = (b_1, \ldots, b_{r+s}) \oplus [0; 0, p-r-s] \oplus [-r+s; 0, q],$$

where

$$(4.4.1)\qquad \begin{cases} b_i := \lambda_i + \dfrac{-p+q-r+s-1}{2} + i & (1 \le i \le r) \\ b_{r+i} := \lambda_{r+i} + \dfrac{p-q-r-s-1}{2} + i & (1 \le i \le s) \end{cases}$$

Proposition 4.4.2. *Retain notations as above. Let $\lambda_i \in \mathbb{Z} + Q\,(1 \le i \le r+s)$, $\delta \in \mathbb{Z}^{r+s}$. Recall that we defined $Q = \frac{1}{2}(p+q-r-s-1)$. Assume that*

$$(4.4.3)(\mathrm{a})\qquad Q \ge \lambda_1 > \lambda_2 > \cdots > \lambda_r \ge -Q,$$

$$(4.4.3)(\mathrm{b})\qquad Q \ge \lambda_{r+1} > \lambda_{r+2} > \cdots > \lambda_{r+s} \ge -Q,$$

$$(4.4.3)(\mathrm{c})\qquad \delta_1 \ge \delta_2 \ge \cdots \ge \delta_r \ge 0 \ge \delta_{r+1} \ge \delta_{r+2} \ge \cdots \ge \delta_{r+s}.$$

We define $y^{(1)} \equiv y^{(1)}(\lambda) = (y_1^{(1)}, \dots, y_r^{(1)}) \in \mathbb{Z}^r$, $y^{(2)} \equiv y^{(2)}(\lambda) = (y_1^{(2)}, \dots, y_s^{(2)}) \in \mathbb{Z}^s$, $z^{(1)} \equiv z^{(1)}(\delta) = (z_1^{(1)}, \dots, z_r^{(1)}) \in \mathbb{Z}^r$ and $z^{(2)} \equiv z^{(2)}(\delta) = (z_1^{(2)}, \dots, z_s^{(2)}) \in \mathbb{Z}^s$ by $y_i^{(1)} := i - b_i = -\lambda_i - \frac{1}{2}(-p+q-r+s-1)\,(1 \le i \le r)$, $y_i^{(2)} := i - b_{r+i} = -\lambda_{r+i} - \frac{1}{2}(p-q-r-s-1)\,(1 \le i \le s)$, $z_j^{(1)} := \delta_j - j\,(1 \le j \le r)$ and $z_j^{(2)} := \delta_{r+j} - j\,(1 \le j \le s)$. Set

$$(4.4.4)(\mathrm{a})\qquad \begin{cases} k := \max[\{i; 1 \le i \le r,\, b_i \ge 0\} \cup \{0\}], \\ l := \max[\{i; 1 \le i \le s,\, 0 \ge b_{r+s+1-i}\} \cup \{0\}], \end{cases}$$

$$(4.4.4)(\mathrm{b})\qquad \tilde{\mu_\lambda} := (b_1, \dots, b_k, 0, \dots, 0, b_{r+s-l+1}, \dots, b_{r+s}, 0, \dots, 0) \in \mathfrak{t}^*$$

Then the following holds:

1) $M(\mathfrak{q}, \lambda, \delta) = F(q, r; y^{(1)}(\lambda), z^{(1)}(\delta)) \cdot F(q, s; -y^{(2)}(\lambda), -z^{(2)}(\delta))$.

2) *Assume moreover that $\sum_{i=1}^r \delta_i = \sum_{i=1}^k b_i$ and that $\sum_{i=1}^s \delta_{r+i} = \sum_{i=1}^l b_{r+s-l+i}$.*

 a) *If $\delta \ne \tilde{\mu_\lambda}$, then $M(\mathfrak{q}, \lambda, \delta) = 0$.*

 b) *If $\delta = \tilde{\mu_\lambda}$, then*

$$M(\mathfrak{q}, \lambda, \delta) = d(q, r-k; y_{k+1}^{(1)}, \dots, y_r^{(1)}) \cdot d'(q, s-l; -y_1^{(2)}, \dots, -y_{s-l}^{(2)}).$$

3) $M(\mathfrak{q}, \lambda, \delta) = 0$ *if $\sum_{i=1}^r \delta_i < \sum_{i=1}^k b_i$ or if $\sum_{i=1}^s \delta_{r+i} > \sum_{i=1}^l b_{r+s-l+i}$.*

Remark 4.4.5. In the above Proposition, the assumption $Q \geq \lambda_1$ in (4.4.3)(a) (respectively, $\lambda_{r+s} \geq -Q$ in (4.4.3)(b)) can be relaxed to $Q \geq \lambda_{k+1}$ (respectively, $\lambda_{r+s-l} \geq -Q$) in (2) and can be dropped in (3).

The argument parallels Proposition(4.3.2) in the quarternionic case; we will explicitly give only the necessary steps in §4.8. We observe that $r - q > 0$ iff $r - q > k$ under the assumption (4.4.3)(a) and that $s - q > 0$ iff $s - q > l$ under the assumption (4.4.3)(b). Indeed, $b_{r-q} = \lambda_{r-q} - Q - 1$ when $r > q$ and $b_{r+q+1} = \lambda_{r+q+1} + Q + 1$ when $s > q$.

Since $y_i^{(1)}(\lambda) = i - b_i \geq 1\,(k+1 \leq i \leq r)$, $-y_i^{(2)}(\lambda) = -i + b_{r+i} \geq -(s-l)\,(1 \leq i \leq s-l)$, combining Proposition(4.4.2) with Lemma(4.2.5),(4.2.6) and (4.2.8), we have

Corollary 4.4.6. *Retain the same notations in Proposition*(4.4.2). *If* $(\lambda_1, \ldots, \lambda_{r+s}) \in \mathbb{Z}^{r+s} + [Q; 0, r+s]$ *satisfies*

$$\lambda_1 > \lambda_2 > \cdots > \lambda_r \geq -Q,$$

$$Q \geq \lambda_{r+1} > \lambda_{r+2} > \cdots > \lambda_{r+s},$$

and if $\mathcal{R}_{\mathfrak{q}}^{S-i}(\mathbb{C}_\lambda) = 0$ *for* $i \neq 0$, *then the following two conditions are equivalent:*

a) $\mathcal{R}_{\mathfrak{q}}^{S}(\mathbb{C}_\lambda) = 0$.

b) $q < r$ *and* $\lambda_{r-q} \leq Q$ or $q < s$ *and* $\lambda_{r+q+1} \geq -Q$.

4.5. result in real case

Retain notations in §4.1-2 and §2.5. Note that $\mathfrak{t}^c = \mathfrak{h}$ iff pq is even. Set $p' := [\frac{p}{2}]$ and $q' := [\frac{q}{2}]$. We have

$$\rho_c = [\frac{p}{2} - 1; -1, p'] \oplus [\frac{q}{2} - 1; -1, q'],$$

$$\rho(\mathfrak{u})_{|\mathfrak{t}^c} = [\frac{p+q}{2} - 1; -1, r] \oplus [0; 0, p' + q' - r],$$

$$\rho(\mathfrak{u} \cap \mathfrak{k}) = [\frac{p}{2} - 1; -1, r] \oplus [0; 0, p' + q' - r],$$

Thus, $\rho(\mathfrak{u}) - 2\rho(\mathfrak{u}\cap\mathfrak{k})_{|\mathfrak{t}^c} = [\frac{-p+q}{2}+1;1,r] \oplus [0;0,p'+q'-r]$.

For $\lambda := (\lambda_1,\dots,\lambda_r,0,\dots,0)$, $\lambda' := \lambda - 2\lambda_r f_r \in (\mathfrak{t}^c)^*(\subset -\mathfrak{h}^*)$, we set

$$\mu_\lambda : = \lambda + \rho(\mathfrak{u}) - 2\rho(\mathfrak{u}\cap\mathfrak{k}) = (b_1,\dots,b_r,0,\dots,0) \in (\mathfrak{t}^c)^*,$$

$$\mu'_{\lambda'} : = \lambda' + \rho(\mathfrak{u}') - 2\rho(\mathfrak{u}'\cap\mathfrak{k}) = (b'_1,\dots,b'_r,0,\dots,0) \in (\mathfrak{t}^c)^*,$$

where

$$\text{(4.5.1)} \qquad \begin{cases} b_i : = \lambda_i + \dfrac{-p+q}{2} + i & (1 \le i \le r), \\ b_i = b'_i \ (1 \le i \le r-1), \quad b_r = -b'_r. \end{cases}$$

Proposition 4.5.2. *Retain notations as above. Let $\lambda_i \in \mathbb{Z} + Q\,(1 \le i \le r)$, $\delta \in \mathbb{Z}^r$. Recall that $Q = \frac{1}{2}(p+q) - r - 1$. Assume that*

(4.5.3)(a) $\qquad Q \ge \lambda_1 > \lambda_2 > \cdots > \lambda_r \ge -Q,$

(4.5.3)(b) $\qquad \delta_1 \ge \delta_2 \ge \cdots \ge \delta_r \ge 0 \qquad (p \ne 2r)$

(4.5.3)(b′) $\qquad \delta_1 \ge \delta_2 \ge \cdots \ge \delta_{r-1} \ge |\delta_r| \qquad (p = 2r)$

We define $y \equiv y(\lambda) = (y_1,\dots,y_r) \in \mathbb{Z}^r$, $z \equiv z(\delta) = (z_1,\dots,z_r) \in \mathbb{Z}^r$ by $y_i := i - b_i = -\lambda_i - \frac{p-q}{2}\,(1 \le i \le r)$, $z_j := \delta_j - j\,(1 \le j \le r)$. Set

(4.5.4)(a) $\qquad k := \max[\{i; 1 \le i \le r,\ b_i \ge 0\} \cup \{0\}].$

(4.5.4)(b) $\qquad \tilde{\mu_\lambda} := (b_1,\dots,b_k,0,\dots,0).$

Then the following holds:

1) *$M(\mathfrak{q},\lambda,\delta) = F(q,r;y(\lambda),z(\delta))$.*

2) *Assume moreover that $\sum_{i=1}^r \delta_i = \sum_{i=1}^k b_i$. Then*

a) *If* $\delta \neq \tilde{\mu_\lambda}$, *then* $M(\mathfrak{q}, \lambda, \delta) = 0$.

b) *If* $\delta = \tilde{\mu_\lambda}$, *then* $M(\mathfrak{q}, \lambda, \delta) = d(q, r-k; y_{k+1}(\lambda), \dots, y_r(\lambda))$.

3) $M(\mathfrak{q}, \lambda, \delta) = 0$ *if* $\sum_{i=1}^{r} \delta_i < \sum_{i=1}^{k} b_i$.

Similar statements are valid for λ'.

Corollary 4.5.5. *Retain the same notations in Proposition(4.5.2). If* $(\lambda_1, \dots, \lambda_r) \in \mathbb{Z}^r + [Q; 0, r]$ *satisfies*

$$\lambda_1 > \lambda_2 > \cdots > \lambda_r \geq -Q,$$

and if $\mathcal{R}_{\mathfrak{q}}^{S-i}(\mathbb{C}_\lambda) = \mathcal{R}_{\mathfrak{q}'}^{S-i}(\mathbb{C}_{\lambda'}) = 0$ *for* $i \neq 0$, *then the following three conditions are equivalent:*

a) $\mathcal{R}_{\mathfrak{q}}^{S}(\mathbb{C}_\lambda) = 0$.

a′) $\mathcal{R}_{\mathfrak{q}'}^{S}(\mathbb{C}_{\lambda'}) = 0$.

b) $q < r$ *and* $\lambda_{r-q} \leq Q$.

4.6. some auxiliary lemmas

In this subsection, we collect elementary lemmas used in the proof of §4.3-5. Our strategy is to reduce an alternating sum in a generalized Blattner formula to the determinant of some matrix. The main point is that only a (possibly small) symmetric group contained in a Weyl group appears in the actual calculation. In particular the elements in a Weyl group which reverse the signature of coordinates do not appear under suitable inequalities of vectors. This is what we describe here.

Fix p, $r \in \mathbb{N}$ with $1 \leq r \leq p$.

Lemma 4.6.1. *Let* $a = (a_1, \dots, a_r, 0, \dots, 0) \in \mathbb{Z}^p$. *Assume that*

$$a_j \geq -2p + 2r - 1 \text{ for any } j \quad (1 \leq j \leq r). \tag{4.6.2}$$

If

$$w\cdot(\delta+\langle p\rangle)-\langle p\rangle=a \tag{4.6.3}$$

for some $w\in\mathfrak{S}_p\ltimes{\mathbb{Z}_2}^p$, $\delta=(\delta_1,\dots,\delta_p)\in\mathbb{N}^p$, *then we have*

$$\begin{cases} w\in\mathfrak{S}_r\times 1_{p-r}\,(\subset\mathfrak{S}_r\times\mathfrak{S}_{p-r}\subset\mathfrak{S}_p\ltimes{\mathbb{Z}_2}^p),\\ \delta_j=0 \qquad (r+1\le j\le p).\end{cases}$$

Proof. The equation(4.6.3) is written as

$$\begin{aligned} &w\cdot(\delta_1+p,\delta_2+p-1,\dots,\delta_p+1)\\ &\qquad=(a_1+p,a_2+p-1,\dots,a_r+p+1-r,p-r,p-r-1,\dots,1).\end{aligned} \tag{4.6.3$'$}$$

Therefore there are at least $p-r$ elements among $\{\delta_j+p+1-j;1\le j\le p\}$ whose absolute values are not greater than $p-r$. Since $\delta_j\ge 0$, this implies $w\in(\mathfrak{S}_r\ltimes{\mathbb{Z}_2}^r)\times 1_{p-r}$ and $\delta_j=0$ $(r+1\le j\le p)$.

Now we must show that w does not reverse any signs. The condition (4.6.2) assures that $a_j+p+1-j\ge-(p-r)$, while $\delta_j+p+1-j\ge p+1-j\ge p+1-r$ for $1\le j\le r$. Hence $a_j+p+1-j$ must be positive and $w\in\mathfrak{S}_r\times 1_{p-r}$, which completes the proof. Note that we have also proved at the same time that $a_j+p+1-j\ge p+1-r$, that is, $a_j\ge -r+j$ $(1\le j\le r)$. □

Lemma 4.6.4. *Let* $b=(b_1,\dots,b_r)\in\mathbb{Z}^r$ *satisfy*

$$b_1\ge b_2\ge\cdots\ge b_k\ge 0, \tag{4.6.5)(a}$$

$$0>b_j \quad (k+1\le j\le r), \tag{4.6.5)(b}$$

for some k $(0\le k\le r)$. *Set*

$$n:=-\sum_{j=k+1}^{r}b_j\in\mathbb{N}. \tag{4.6.6}$$

Assume that there are $w \in \mathfrak{S}_r$, $c = (c_1, \dots, c_r) \in \mathbb{Z}^r$, $\delta = (\delta_1, \dots, \delta_r) \in \mathbb{Z}^r$ *such that*

$$w \cdot (\delta + \langle r \rangle) - \langle r \rangle = b + c, \tag{4.6.7}$$

$$c_j \geq 0 \qquad (1 \leq j \leq k), \tag{4.6.8)(a}$$

$$\delta_1 \geq \delta_2 \geq \cdots \geq \delta_r \geq 0 \tag{4.6.8)(b}$$

Then the following holds.

1) $\sum_{j=1}^r c_j = \sum_{j=1}^r (\delta_j - b_j) \geq n$.

2) *If* $\sum_{j=1}^r c_j = n$ *or equivalently if* $\sum_{j=1}^r \delta_j = \sum_{j=1}^k b_j$, *then*

$$\begin{cases} w \in 1_k \times \mathfrak{S}_{r-k}, \\ \delta_j = b_j \ (1 \leq j \leq k), \quad \delta_j = 0 \ (k+1 \leq j \leq r), \\ c_j = 0 \ (1 \leq j \leq k), \quad c_{k+j} = -b_{k+j} + j - w \cdot j \ (1 \leq j \leq r-k). \end{cases}$$

3) *If* $\sum_{j=1}^r c_j > n$, *then* $\sum_{j=1}^r \delta_j > \sum_{j=1}^k b_j$. *In particular, we have*

$$\delta \neq (b_1, \dots, b_k, 0, \dots, 0).$$

Proof. From definition(4.6.6) of n and the equation(4.6.7), we have

$$\sum_{j=1}^r \delta_j = \sum_{j=1}^r (b_j + c_j) = \sum_{j=1}^k b_j + \sum_{j=1}^r c_j - n. \tag{4.6.9}$$

Using inequalities (4.6.8)(a),(b), we have

$$\sum_{j=1}^k b_j \leq \sum_{j=1}^k (b_j + c_j) \leq \sum_{j=1}^k \delta_j \leq \sum_{j=1}^r \delta_j. \tag{4.6.10}$$

Here the middle inequality is because $b + c - \delta = w \cdot (\delta + \langle r \rangle) - (\delta + \langle r \rangle)$ and because the components of $\delta + \langle r \rangle$ are strictly decreasing.

The statement (3) is clear from (4.6.9). Substitution of (4.6.9) into (4.6.10) gives $\sum_{j=1}^{r} c_j \geq n$, namely, (1).

Finally assume $\sum_{j=1}^{r} c_j = n$. Then all the terms in (4.6.10) must be equal, which implies

$$\begin{cases} w \in \mathfrak{S}_k \times \mathfrak{S}_{r-k}, \\ \delta_j = 0 \qquad (k+1 \leq j \leq r), \\ c_j = 0 \qquad (1 \leq j \leq k). \end{cases}$$

Then both of the first k coordinates of $\delta + \langle r \rangle$ and $b + c + \langle r \rangle$ are strictly decreasing, so that $w \in 1_k \times \mathfrak{S}_{r-k}$. Now the remaining assertions in (2) are clear. □

The following result is a direct consequence of Lemma(4.6.1) and Lemma(4.6.4).

Proposition 4.6.11. *Let $b = (b_1, \ldots, b_r, 0, \ldots, 0) \in \mathbb{Z}^p$. Assume that*

$$\begin{cases} b_1 \geq b_2 \geq \cdots \geq b_k \geq 0 \\ 0 > b_j \geq -2p + 2r - 1 \quad (k+1 \leq j \leq r), \end{cases}$$

for some k $(0 \leq k \leq r)$. Suppose that there are $w \in \mathfrak{S}_p \ltimes {\mathbb{Z}_2}^p$, $c = (c_1, \ldots, c_r, 0, \ldots, 0) \in \mathbb{Z}^p$ and $\delta = (\delta_1, \ldots, \delta_p) \in \mathbb{Z}^p$ such that

$$\begin{cases} w \cdot (\delta + \langle p \rangle) - \langle p \rangle = b + c, \\ c_j \geq 0 \qquad (1 \leq j \leq r), \\ \delta_1 \geq \delta_2 \geq \cdots \geq \delta_p \geq 0. \end{cases}$$

Set $n := -\sum_{j=k+1}^{r} b_j$. Then the following holds.

1) $\sum_{j=1}^{r} c_j \geq n$, $\sum_{j=1}^{r} \delta_j \geq \sum_{j=1}^{k} b_j$, $w \in \mathfrak{S}_r \times 1_{p-r}$, *and* $\delta_j = 0 \quad (r+1 \leq j \leq p)$.

2) *If* $\sum_{j=1}^{r} c_j = n$ *or equivalently if* $\sum_{j=1}^{r} \delta_j = \sum_{j=1}^{k} b_j$, *then,*

$$\begin{cases} w \in 1_k \times \mathfrak{S}_{r-k}, \\ \delta_j = b_j \ (1 \le j \le k), \quad \delta_j = 0 \ (k+1 \le j \le r), \\ c_j = 0 \ (1 \le j \le k), \quad c_{k+j} = -b_{k+j} + j - w \cdot j \ (1 \le j \le r-k). \end{cases}$$

3) *If* $\sum_{j=1}^{r} c_j > n$, *then* $\sum_{j=1}^{r} \delta_j > \sum_{j=1}^{k} b_j$. *In particular,*

$$\delta \neq (b_1, \ldots, b_k, 0, \ldots, 0).$$

4.7. proof for quarternionic case

In this subsection, we complete the proof of Proposition(4.3.2). Suppose we are in the setting of §4.3. First recall (cf. §2.1) that the root system of $\mathfrak{u}$ for $\mathfrak{t}^c = \mathfrak{h}$ is given by

$$\Delta(\mathfrak{u} \cap \mathfrak{p}, \mathfrak{h}) := \{f_i \pm f_j \,;\, 1 \le i \le r, \ p+1 \le j \le p+q\},$$

and that $L \cap K \simeq \mathbb{T}^r \times Sp(p-r) \times Sp(q)$ acts on $\mathfrak{u} \cap \mathfrak{p} \simeq \mathbb{C}^{2rq}$ by

$$\bigoplus_{i=1}^{r} (F(\mathbb{T}^r, f_i) \boxtimes \mathbf{1} \boxtimes F(Sp(q), (1, 0, \ldots, 0))).$$

Assume that $\lambda \in \mathbb{Z}^r$ satisfies

$$\lambda_1 > \lambda_2 > \cdots > \lambda_r \ge -Q \equiv -p - q + r.$$

Then $\mu_\lambda \equiv (b_1, \ldots, b_r, 0, \ldots, 0)$ with $b_j \equiv \lambda_j - p + q + j - 1$ satisfies

$$b_1 \ge b_2 \ge \cdots \ge b_r \ge -2p + 2r - 1 \tag{4.7.1}$$

Define $k \, (0 \le k \le r)$ by (4.3.4).

Lemma 4.7.2. *Retain notation as above, let $\lambda \in \mathbb{Z}^r$, $\delta \in \mathbb{Z}^p$ with the same hypotheses on λ and $\delta_1 \geq \cdots \geq \delta_p \geq 0$.*

1) *If $(\delta_{r+1}, \ldots, \delta_p) = (0, \ldots, 0)$, then*

$$m(\mathfrak{q}, \lambda, \delta) = \sum_{w,\{c_i\}} \operatorname{sgn}(w) \prod_{i=1}^{r} \dim S^{c_i}(\mathbb{C}^{2q}). \tag{4.7.3}$$

(See §4.1 for notation.) *Here the sum is taken over the set satisfying*

$$w \in \mathfrak{S}_r, \quad c = (c_1, \ldots, c_r) \in \mathbb{N}^r, \tag{4.7.4)(a}$$

$$w \cdot ((\delta_1, \ldots, \delta_r) + \langle r \rangle) - \langle r \rangle = b + c. \tag{4.7.4)(b}$$

2) *If*

$$(\delta_{r+1}, \ldots, \delta_p) \neq (0, \ldots, 0) \quad \text{or} \quad \sum_{i=1}^{r} \delta_i < \sum_{i=1}^{k} b_i, \tag{4.7.5}$$

then $m(\mathfrak{q}, \lambda, \delta) = 0$.

Proof. As an $L \cap K_1 \simeq \mathbb{T}^r \times Sp(p-r)$-module,

$$S(\mathfrak{u} \cap \mathfrak{p}) \simeq \bigoplus_{c \in \mathbb{N}^r} \left(\prod_{i=1}^{r} \dim S^{c_i}(\mathbb{C}^{2q}). \right) \cdot (c_1, \ldots, c_r) \boxtimes \mathbf{1},$$

where the multiplier in the right hand stands for the multiplicity of each representation and $c = (c_1, \ldots, c_r) \in \mathbb{Z}^r$ denotes an additive character of $\mathbb{T}^r$. Thus for each $\pi_1 \in \widehat{K_1}$,

$$\begin{aligned} &\operatorname{Hom}_{L \cap K_1}(H^j(\mathfrak{u} \cap \mathfrak{k}_1, \pi_1), S(\mathfrak{u} \cap \mathfrak{p}) \otimes \mathbb{C}_{\mu_\lambda}) \\ &\simeq \bigoplus_{c \in \mathbb{N}^r} \left(\prod_{i=1}^{r} \dim S^{c_i}(\mathbb{C}^{2q}) \right) \cdot \operatorname{Hom}_{\mathbb{T}^r \times Sp(p-r)}(H^j(\mathfrak{u} \cap \mathfrak{k}_1, \pi_1), (b+c) \boxtimes \mathbf{1}). \end{aligned} \tag{4.7.6}$$

We will regard c as an element of $\mathbb{N}^p$ by putting $c = (c_1, \ldots, c_r, 0, \ldots, 0)$ without comment. Appealing to Kostant's Borel-Weil theorem

$$\dim \operatorname{Hom}_{\mathbb{T}^r \times Sp(p-r)}(H^j(\mathfrak{u} \cap \mathfrak{k}_1, \pi_1), (b+c) \boxtimes \mathbf{1})$$

is the number of elements w of

$$W_{K_1}^{\mathfrak{l}\cap\mathfrak{k}_1} := \{w \in W(\mathfrak{k}_1);\ \Delta^+(\mathfrak{k}_1) \cap w \cdot \Delta^-(\mathfrak{k}_1) \subset \Delta(\mathfrak{u} \cap \mathfrak{k}_1)\}$$

such that

$$(4.7.7)(a) \qquad l(w) = j,$$

$$(4.7.7)(b) \qquad w \cdot (\delta + \langle p \rangle) - \langle p \rangle = b + c.$$

Assume that (4.7.7)(b) holds. From (4.7.1) and $c_i \geq 0$, we are able to apply Proposition(4.6.11), so that we conclude $w \in \mathfrak{S}_r \times 1_{p-r}$, $\delta_i = 0$ $(r+1 \leq i \leq p)$ and $\sum_{i=1}^{r} \delta_i \geq \sum_{i=1}^{k} b_i$. Thus the statement (2) is clear. Now Part (1) is followed from (4.7.6) since $\mathfrak{S}_r \times 1_{p-r} \subset W_{K_1}^{\mathfrak{l}\cap\mathfrak{k}_1}$. □

Proof of Proposition(4.3.2). First notice that we have

$$(4.7.8) \qquad M(\mathfrak{q}, \lambda, \delta) = m(\mathfrak{q}, \lambda, \delta \oplus [0; 0, p-r]),$$

from Lemma(4.7.2). Therefore the third statement is plain in view of Lemma(4.7.2)(2).

1) . Fix $w \in \mathfrak{S}_r$ and define $c \in \mathbb{Z}^r$ by the formula(4.7.4)(b). Then putting $j := w \cdot i$ $(1 \leq i \leq r)$, we have

$$\begin{aligned} c_i &= -b_i + (\delta_j + r - j + 1) - (r - i + 1) \\ &= -b_i + (\delta_j - j) + i \\ &= -\lambda_i + p - q + 1 + \delta_j - j \\ &\geq -(p + q - r) + p - q + 1 - r \\ &= 1 - 2q, \end{aligned} \qquad (4.7.9)$$

for any i $(1 \leq i \leq r)$. In the third inequality we used (4.3.3)(a). Then this inequality(4.7.9) assures the condition for (4.2.4) so that we have

$$a(c_i, 2q) = \dim S^{c_i}(\mathbb{C}^{2q}).$$

Therefore, combining (4.7.3) with (4.7.8), we get

$$M(\mathfrak{q},\lambda,\delta) = \sum_{w\in\mathfrak{S}_r} \operatorname{sgn}(w)\left(\prod_{i=1}^{r} a(i-b_i+\delta_{w\cdot i}-w\cdot i,2q)\right) \tag{4.7.10}$$

$$= F(2q,r;y(\lambda),z(\delta)) \qquad (\text{Definition}(4.2.2)(\text{a})).$$

2) Assume $\sum_{i=1}^{r}\delta_i = \sum_{i=1}^{k} b_i$. From Lemma(4.6.4)(2), each term in the right hand side of (4.7.3) is non-zero only if $\delta = (b_1,\dots,b_k,0,\dots,0)$ and $w \in 1_k \times \mathfrak{S}_{r-k}$. From (4.7.8), the first statement follows. If $\delta = (b_1,\dots,b_k,0,\dots,0)$, then we have

$$M(\mathfrak{q},\lambda,\delta) = \sum_{w\in 1_k\times\mathfrak{S}_{r-k}} \operatorname{sgn}(w)\left(\prod_{i=k+1}^{r} a(i-b_i-w\cdot i,2q)\right)$$

$$= d(2q,r-k;y_{k+1}(\lambda),\dots,y_r(\lambda)),$$

which is the desired result. □

We remark that in the proof of the second statement $i-b_i-w\cdot i \geq 1-2q$ $(k+1\leq i\leq r)$ if $Q \geq \lambda_{k+1}$. In this way Remark(4.3.5) is justified.

4.8. proof for complex case

Set $v \in \mathbb{C}^p(\simeq (\mathfrak{t}_1^c)^*)$ by

$$v := [\frac{p-1}{2};-1,r] \oplus [\frac{-p+2s-1}{2};-1,s] \oplus [\frac{p-2r-1}{2};-1,p-r-s]$$
$$= [p;-1,r] \oplus [s;-1,s] \oplus [p-r;-1,p-r-s] + [\frac{-p-1}{2};0,p].$$

We need first the next Lemma as a slight variant of Proposition(4.6.11).

Lemma 4.8.1. *Let $b = (b_1,\dots,b_r,b_{r+1},\dots,b_{r+s},0,\dots,0) \in \mathbb{Z}^p$. Assume that*

$$b_1 \geq b_2 \geq \cdots \geq b_k \geq 0, \tag{4.8.2)(a}$$

$$0 > b_i \geq -p+s+i \qquad (k+1\leq i\leq r), \tag{4.8.2)(b}$$

$$p-r-s-1+i \geq b_{r+i} > 0 \qquad (1\leq i\leq s-l), \tag{4.8.2)(c}$$

$$0 \geq b_{r+s+1-l} \geq \cdots \geq b_{r+s}, \tag{4.8.2)(d}$$

for some k $(0 \leq k \leq r)$ *and* l $(0 \leq l \leq s)$. *Suppose that there are* $w \in \mathfrak{S}_p$, $c = (c_1, \dots, c_{r+s}, 0, \dots, 0) \in \mathbb{Z}^p$, $\delta = (\delta_1, \dots, \delta_p) \in \mathbb{Z}^p$ *such that*

(4.8.3)(a) $$w \cdot (\delta + v) - v = b + c,$$

(4.8.3)(b) $$c_i \geq 0\ (1 \leq i \leq r), \quad c_i \leq 0\ (r+1 \leq i \leq r+s),$$

(4.8.3)(c)
$$\delta_1 \geq \delta_2 \geq \cdots \geq \delta_r \geq \delta_{r+s+1} \geq \delta_{r+s+2} \geq \cdots \geq \delta_p \geq \delta_{r+1} \geq \delta_{r+2} \geq \cdots \geq \delta_{r+s}.$$

Set $n := -\sum_{i=k+1}^{r} b_i$, $\quad m := \sum_{i=1}^{s-l} b_{r+i} \in \mathbb{N}$. *Then the following holds:*

1) $w \in \mathfrak{S}_r \times \mathfrak{S}_s \times 1_{p-r-s}$, $\delta_i = 0\ (r+s+1 \leq i \leq p)$, $\sum_{i=1}^{r} c_i \geq n$, *and* $\sum_{i=1}^{s} c_{r+i} \leq -m$.

2) *If*

(4.8.4) $$\sum_{i=1}^{r} c_i = n \quad \text{and} \quad \sum_{i=1}^{s} c_{r+i} = -m,$$

then,

$$\begin{cases} w \in 1_k \times \mathfrak{S}_{r-k} \times \mathfrak{S}_{s-l} \times 1_l \times 1_{p-r-s}, \\ \delta_i = b_i \quad (1 \leq i \leq k,\ r+s-l+1 \leq i \leq r+s), \\ \delta_i = 0 \quad (k+1 \leq i \leq r+s-l,\ r+s+1 \leq i \leq p) \\ c_i = 0 \quad (1 \leq i \leq k,\ r+s-l+1 \leq i \leq r+s), \\ c_{k+i} = -b_{k+i} + i - w' \cdot i \quad (1 \leq i \leq r-k), \\ c_{r+i} = -b_{r+i} + i - w'' \cdot i \quad (1 \leq i \leq s-l). \end{cases}$$

Here $w' \in \mathfrak{S}_{r-k}$ *and* $w'' \in \mathfrak{S}_{s-l}$ *are the obvious restrictions of* w.

3) *If* $\sum_{i=1}^{r} c_i > n$, *then* $\sum_{i=1}^{r} \delta_i > \sum_{i=1}^{k} b_i$. *If* $\sum_{i=1}^{s} c_{r+i} < -m$, *then* $\sum_{i=1}^{s} \delta_{r+i} > \sum_{i=1}^{l} b_{r+s-l+i}$. *In particular, if* (4.8.4) *is not satisfied, then*

$$\delta \neq (b_1, \dots, b_k) \oplus [0; 0, r+s-k-l] \oplus (b_{r+s+1-l}, \dots, b_{r+s}) \oplus [0; 0, p-r-s].$$

Proof. We may replace v by $[p;-1,r]\oplus[s;-1,s]\oplus[p-r;-1,p-r-s]$ without changing notations. The condition(4.8.2)(b)(c) and (4.8.3)(b) assure

$$\begin{aligned}
(b+c+v)_i &\equiv b_i+c_i+p+1-i && \geq s+1\ (1\leq i\leq r)\\
(b+c+v)_{r+i} &\equiv b_{r+i}+c_{r+i}+s+1-i \leq p-r\ (1\leq i\leq s)\\
\{(b+c+v)_{r+s+i}\,;\,1\leq i\leq p-r-s\} &= \{s+1,s+2,\dots,p-r\}
\end{aligned}$$

From (4.8.3)(a)(c), $b+c+v$ consists of distinct entries, which implies $w\in\mathfrak{S}_r\times\mathfrak{S}_s\times 1_{p-r-s},\ \ \delta_i=0\ (r+s+1\leq i\leq p)$.

Now we apply (4.6.4) to the first r block (respectively the second s block) and then the lemma follows. □

Suppose we are in the setting of §4.4. Recall that the root system of $\mathfrak{u}$ for $\mathfrak{t}^c=\mathfrak{h}$ is given by (see §2.3)

$$\begin{aligned}
\Delta(\mathfrak{u}\cap\mathfrak{p},\mathfrak{h}) := \{f_i-f_j\,;\,1\leq i\leq r,\ p+1\leq j\leq p+q\}\\
\cup\ \{-f_i+f_j\,;\,r+1\leq i\leq r+s,\ p+1\leq j\leq p+q\}.
\end{aligned}$$

and that $L\cap K\simeq\mathbb{T}^{r+s}\times U(p-r-s)\times U(q)$ acts on $\mathfrak{u}\cap\mathfrak{p}\simeq\mathbb{C}^{(r+s)q}$ by

$$\bigoplus_{i=1}^{r}\left(F(\mathbb{T}^{r+s},f_i)\boxtimes\mathbf{1}\boxtimes F(U(q),(-1,0,\dots,0))\right)$$
$$\oplus\bigoplus_{i=1}^{s}\left(F(\mathbb{T}^{r+s},-f_{r+i})\boxtimes\mathbf{1}\boxtimes F(U(q),(1,0,\dots,0))\right).$$

Assume that

$$\begin{cases}
\lambda_i\in\mathbb{Z}+Q,\\
\lambda_1>\lambda_2>\cdots>\lambda_r\geq -Q.\\
Q\geq\lambda_{r+1}>\lambda_{r+2}>\cdots>\lambda_{r+s}
\end{cases}$$

Then $\mu_\lambda = (b_1, \dots, b_{r+s}, 0, \dots, 0)$ (Definition (4.4.1)) satisfies

$$\begin{cases} b_1 \geq b_2 \geq \cdots \geq b_r \geq -p+s+r, \\ p-r-s \geq b_{r+1} \geq b_{r+2} \geq \cdots \geq b_{r+s}. \end{cases}$$

Now the proof of Proposition(4.4.2) is done in the same way as the previous section based on the following lemma (cf. Lemma(4.7.2)). We omit the proof.

Lemma 4.8.5. *Retain notation as above, let $\lambda \in \mathbb{Z}^{r+s} + [Q; 0, r+s]$, $\delta \in \mathbb{Z}^p$ with the same hypotheses on λ and $\delta_1 \geq \delta_2 \geq \cdots \geq \delta_r \geq \delta_{r+s+1} \geq \delta_{r+s+2} \geq \cdots \geq \delta_p \geq \delta_{r+1} \geq \delta_{r+2} \geq \cdots \geq \delta_{r+s}$. Then*

1) *If $\delta_{r+s+1} = \cdots = \delta_p = 0$, then*

$$(4.8.6) \quad m(\mathfrak{q}, \lambda, \delta) = \left(\sum_{w', \{c_i\}} \operatorname{sgn}(w') \prod_{i=1}^{r} \dim S^{c_i}(\mathbb{C}^q) \right) \times \left(\sum_{w'', \{c_{r+i}\}} \operatorname{sgn}(w'') \prod_{i=1}^{s} \dim S^{-c_{r+i}}(\mathbb{C}^q) \right)$$

Here the sums are taken over the sets satisfying

$$(4.8.7)(\mathrm{a}) \qquad w' \in \mathfrak{S}_r, \quad c_i \in \mathbb{N}\,(1 \leq i \leq r),$$

$$(4.8.7)(\mathrm{b}) \qquad w' \cdot ((\delta_1, \dots, \delta_r) + \langle r \rangle) - \langle r \rangle = (b_1 + c_1, \dots, b_r + c_r),$$

$$(4.8.8)(\mathrm{a}) \qquad w'' \in \mathfrak{S}_s, \quad c_{r+i} \in -\mathbb{N}\,(1 \leq i \leq s),$$

(4.8.8)(b)

$$w'' \cdot (\delta_{r+1}, \dots, \delta_{r+s}) + \langle s \rangle) - \langle s \rangle = (b_{r+1} + c_{r+1}, \dots, b_{r+s} + c_{r+s}),$$

respectively.

2) *If* $(\delta_{r+s+1},\dots,\delta_p) \neq (0,\dots,0)$ *or* $\sum_{i=1}^{r}\delta_i < \sum_{i=1}^{k} b_i$ *or* $\sum_{i=1}^{s}\delta_{r+i} > \sum_{i=1}^{l} b_{r+s-l+i}$, *then* $m(\mathfrak{q},\lambda,\delta) = 0$.

4.9. proof for real case

$L \cap K \simeq \mathbb{T}^r \times SO(p-r) \times SO(q)$ acts on $\mathfrak{u} \cap \mathfrak{p} \simeq \mathbb{C}^{rq}$ by

$$\bigoplus_{i=1}^{r} \left(F(\mathbb{T}^r, f_i) \boxtimes \mathbf{1} \boxtimes F(SO(q), (-1,0,\dots,0))\right).$$

The proof of Proposition(4.5.2) reduces to the same calculation for the quarternionic case, owing to the following

Lemma 4.9.1. *Let* $a = (a_1,\dots,a_r,0,\dots,0) \in \mathbb{Z}^{p'}$. *Assume that*

$$a_i \geq -p + 2r + 1 \text{ for any } i \quad (1 \leq i \leq r). \tag{4.9.2}$$

If

$$w \cdot (\delta + [\frac{p}{2} - 1; -1, p']) - [\frac{p}{2} - 1; -1, p'] = a$$

for some $w \in W(SO(p))$, $\delta = (\delta_1,\dots,\delta_{p'}) \in \mathbb{N}^{p'}$ *such that* $\delta_1 \geq \cdots \geq \delta_{p'} \geq 0$ *when* p *is odd,* $\delta_1 \geq \cdots \geq \delta_{p'-1} \geq |\delta_{p'}|$ *when* p *is even, then we have* $w \in \mathfrak{S}_r \times 1_{p'-r}$ *and* $\delta_i = 0$ $(r+1 \leq i \leq p')$.

The proof of this Lemma is similar to that of Lemma(4.6.1) and so we omit it. The condition(4.9.2) guarantees $a_i + \frac{p}{2} - i \geq -(\frac{p}{2} - r - 1)$ $(1 \leq i \leq r)$ and is satisfied by putting $a_i = b_i + c_i$ (notation(4.5.1)) when $\lambda_i \geq -Q$ and $c_i \in \mathbb{N}$.

§5. An alternative proof of the sufficiency for $\mathcal{R}_{\mathfrak{q}}^{S}(\mathbb{C}_\lambda) \neq 0$

In studying our discrete series for G/H_2, we are obliged to deal with the parameters outside the fair range. Usually, we move from good parameters toward bad (e.g. fair or non-fair) ones. But this is a *negative* way of thinking. In this section, a *positive* application of bad parameters is offered: making use of direct information for possibly non-fair parameters.

Retain notation in §1. Suppose that $\mathbb{C}_\lambda$ is in the fair range. Then $(\mathfrak{g}, K)$-modules $\mathcal{R}_{\mathfrak{q}}^{j}(\mathbb{C}_\lambda)$ vanish except in the single degree $j = S$, but the remaining module may also vanish as we saw in the previous section. A recent (unpublished) method (cf. §0) to check the sufficiency for $\mathcal{R}_{\mathfrak{q}}^{S}(\mathbb{C}_\lambda) \neq 0$ due to Matsuki and Oshima is based on a computation of very special 'small' K-types case-by-case. They explained to me that it sometimes requires a long calculation and some preparations on finite dimensional representations and that the special case of this section (in particular $\mathbb{F} = \mathbb{H}$ case) is extremely tedious because of a non-triviality of a partition function for $\mathfrak{u} \cap \mathfrak{p}$ as well as a complicated polarization of a parabolic subalgebra.

In the present case $\mathbb{C}_\lambda$ is in the fair range. Nevertheless, treating the parameters *outside* the fair range at the same time leads us to another simple proof of the sufficient condition for $\mathcal{R}_{\mathfrak{q}}^{S}(\mathbb{C}_\lambda) \neq 0$.

Here is our idea:

i) *nice information may be buried outside the fair range.*

ii) *the modules with small (but not necessarily fair) parameters form a closed universe.*

Actually we make use of the difference between the fair range and the condition that μ_λ is $\Delta^+(\mathfrak{k}, \mathfrak{t}^c)$ dominant when λ is small.

5.1. theorem: sufficient condition for $\mathcal{R}_{\mathfrak{q}}^{S}(\mathbb{C}_\lambda) \neq 0$

Let $G = Sp(p,q)$ and $\mathfrak{h}_0$ be a fundamental Cartan subalgebra of $\mathfrak{g}_0$. Take bases of $\mathfrak{h}$ and $\mathfrak{h}^*$ as in §2.1. Fix r and s with $0 < r \leq p$, $0 \leq s \leq q$ and put

(5.1.1)(a) $$\mathfrak{t} := \sum_{i=1}^{r} \mathbb{C}H_i + \sum_{i=1}^{s} \mathbb{C}H_{p+i} \subset \mathfrak{h},$$

(5.1.1)(b) $$Q := p+q-r-s,$$

(5.1.1)(c) $L:$ the centralizer of $\mathfrak{t}$ in G $(\simeq \mathbb{T}^{r+s} \times Sp(p-r, q-s))$.

First of all, let us parametrize θ-stable parabolic subalgebras of $\mathfrak{g}$ with Levi part $\mathfrak{l}$. Let $(m_M, \dots, m_1)$, $(n_N, \dots, n_1)$ be partitions of r and s respectively, that is, $r = \sum_{j=1}^{M} m_j$, $s = \sum_{j=1}^{N} n_j$ with m_j, $n_j \in \mathbb{N}_+$. Assume that $M = N$ or $M = N+1$. Then a θ-stable parabolic subalgebra $\mathfrak{q} = \mathfrak{l} + \mathfrak{u}$ is defined by giving the nilradical $\mathfrak{u}$ so that

(5.1.2) $$\begin{cases} \rho(\mathfrak{u})_{|m_j} = [Q + \sum_{k=1}^{j} m_k + \sum_{k=1}^{j-1} n_k; -1, m_j], \\ \rho(\mathfrak{u})_{|n_j} = [Q + \sum_{k=1}^{j} m_k + \sum_{k=1}^{j} n_k; -1, n_j]. \end{cases}$$

Here we identify $\mathfrak{t}^*$ with $\mathbb{C}^{r+s}$ via the basis $\{H_i\,;\, 1 \leq i \leq r, p+1 \leq i \leq p+s\}$ as usual. We denote by $\rho(\mathfrak{u})_{|m_j}$, $\rho(\mathfrak{u})_{|n_j}$ the obvious restriction of $\rho(\mathfrak{u}) \in \mathfrak{h}^*$ to each subspace $\mathbb{C}^{m_j} \simeq \sum_{i=1}^{m_j} \mathbb{C}H_{m_M + \cdots + m_{j+1} + i}$ $(\subset \mathfrak{h})$, $\mathbb{C}^{n_j} \simeq \sum_{i=1}^{n_j} \mathbb{C}H_{p+n_N+\cdots+n_{j+1}+i}$ $(\subset \mathfrak{h})$, respectively. See §3.5 for the notation of $[k; m, n]$. Conversely, any θ-stable parabolic subalgebra $\mathfrak{q}$ with Levi part $\mathfrak{l}$ is obtained in this manner up to conjugation by an element of K and up to exchange of p by q.

In the setting above, $\mathfrak{q} \cap \mathfrak{k}$ is a parabolic subalgebra of $\mathfrak{k}$ and we have (with similar notations as above):

$$\begin{cases} \rho(\mathfrak{u}\cap\mathfrak{k})_{|m_j} = [p - r + \sum_{k=1}^{j} m_k; -1, m_j], \\ \\ \rho(\mathfrak{u}\cap\mathfrak{k})_{|n_j} = [q - s + \sum_{k=1}^{j} n_k; -1, n_j]. \end{cases}$$

In the present case $\rho(\mathfrak{u}), \rho(\mathfrak{u}\cap\mathfrak{k}) \in \mathfrak{t}^*$ ($\subset \mathfrak{h}^*$ in the notation (1.1.1)). Set

$$\begin{aligned} \mathfrak{t}^*_{\mathbb{Z}} :&= \{\lambda \in \mathfrak{t}^* ; \langle \lambda, \alpha^\vee \rangle \in \mathbb{Z} \quad \text{for all } \alpha \in \Delta(\mathfrak{u}, \mathfrak{h}) \} \\ &= \{\lambda = (\lambda_1, \dots, \lambda_{r+s}) \in \mathfrak{t}^* ; \lambda_i \in \mathbb{Z}\} \\ \mathcal{B} :&= \{\lambda = (\lambda_1, \dots, \lambda_{r+s}) \in \mathfrak{t}^* ; -Q \le \lambda_i \le Q\} \\ \mathcal{F} :&= \{\lambda \in \mathfrak{t}^* ; \langle \lambda, \alpha \rangle > 0 \quad \text{for all } \alpha \in \Delta(\mathfrak{u}, \mathfrak{h}) \} \\ \mathcal{B}_{\mathbb{Z}} :&= \mathcal{B} \cap \mathfrak{t}^*_{\mathbb{Z}}, \qquad \mathcal{F}_{\mathbb{Z}} := \mathcal{F} \cap \mathfrak{t}^*_{\mathbb{Z}}. \end{aligned}$$

Note that a weight $\lambda \in \mathfrak{t}^*$ lifts to a metapletic $(\mathfrak{l}, (L\cap K)^\sim)$-module iff $\lambda \in \mathfrak{t}^*_{\mathbb{Z}}$.

Definition 5.1.3. Suppose we are given $\mathfrak{q} \subset \mathfrak{g}$ as above and $\lambda \in \mathfrak{t}^*$. Set

$$\begin{aligned} r' &:= \#\{i ; 1 \le i \le r, \lambda_i \le Q\}, \\ s' &:= \#\{i ; p+1 \le i \le p+s, \lambda_i \le Q\}, \\ p' &:= p - r + r', \quad q' := q - s + s', \\ \mathfrak{t}''_0 &:= \sum_{i=1}^{r-r'} \mathbb{R}\sqrt{-1}H_i + \sum_{i=1}^{s-s'} \mathbb{R}\sqrt{-1}H_{p+i}, \\ \mathfrak{g}'_0 &:= \text{the semisimple part of the centralizer of } \mathfrak{t}''_0 \text{ in } \mathfrak{g}_0 \\ &\simeq \mathfrak{sp}(p', q') \subset \mathfrak{g}_0, \\ \mathfrak{q}' &:= \mathfrak{g}' \cap \mathfrak{q} \supset \mathfrak{g}. \end{aligned}$$

In extremal cases, $(\mathfrak{g}'_0, \mathfrak{q}') = (\mathfrak{g}_0, \mathfrak{q})$ if $\lambda \in \mathcal{B}$, and $(\mathfrak{g}'_0, \mathfrak{q}') = (\mathfrak{l}_0, \mathfrak{l})$ if $\lambda \in \mathcal{F}$ is sufficiently 'regular'. We shall use similar notations for $(\mathfrak{g}'_0, \mathfrak{q}')$ by adding $'$. For example,

$\mathfrak{t}' = \sum_{i=1}^{r'} \mathbb{C}H_{r-r'+i} + \sum_{i=1}^{s'} \mathbb{C}H_{p+s-s'+i}$, $Q' = (p-r+r') + (q-s+s') - r' - s' = Q$, and etc.

Now we are ready to state the main result in this section.

Theorem 5.1.4. *Let $G = Sp(p,q)$, $\mathfrak{q}$ be a θ-stable parabolic subalgebra of $\mathfrak{g}$ given by (5.1.2), and $\lambda \in \mathcal{F}_{\mathbb{Z}}$. Let $(\mathfrak{g}_0', \mathfrak{q}')$ be associated to $(\mathfrak{g}_0, \mathfrak{q}, \lambda)$ by* Definition(5.1.3). *Then $\mathcal{R}_{\mathfrak{q}}^S(\mathbb{C}_\lambda) \neq 0$ if $-\rho(\mathfrak{u}') + 2\rho(\mathfrak{u}' \cap \mathfrak{k}') \in \mathcal{B}'$.*

We emphasize again that the conclusion was essentially found first by Matsuki and Oshima. In fact, if $\mathcal{R}_{\mathfrak{q}}^S(\mathbb{C}_\lambda)$ is realized as discrete series for a semisimple symmetric space (in this case, the parameter λ is so degenerate as in (2.7.4)(a) and all of m_j, n_j's are even), the above sufficient condition in Theorem(5.1.4) coincides with the necessary one given in [21] after some calculations. We remark that our formulation based on Definition(5.1.3) helps us to understand the somewhat complicated and different description given in [21] in this case.

If $\mathbb{C}_\lambda$ is in the good range with respect to $\mathfrak{q}$, then $\mathfrak{q}' = \mathfrak{l}$ and $-\rho(\mathfrak{u}') + 2\rho(\mathfrak{u}' \cap \mathfrak{k}') = 0 \in \mathcal{B}'$, whence Theorem(5.1.4) implies a well known result $\mathcal{R}_{\mathfrak{q}}^S(\mathbb{C}_\lambda) \neq 0$. The opposite extremal case is essential. Recall that $(\mathfrak{g}_0', \mathfrak{q}') = (\mathfrak{g}_0, \mathfrak{q})$ if $\lambda \in \mathcal{B}$ in Definition(5.1.3). Then the following is a special case of Theorem(5.1.4) by assuming $\lambda \in \mathcal{B}$.

Theorem 5.1.4′ (special case)**.** *With notation as above, if $-\rho(\mathfrak{u}) + 2\rho(\mathfrak{u} \cap \mathfrak{k}) \in \mathcal{B}$, then $\mathcal{R}_{\mathfrak{q}}^S(\mathbb{C}_\lambda) \neq 0$ for all $\lambda \in \mathcal{B} \cap \mathcal{F}_{\mathbb{Z}}$.*

Since the essence of our proof for Theorem(5.1.4) lies in this special case, we shall prove only this case. Indeed, we shall iterate translations among small singular parameters, where the components (strictly) larger than Q are always stable, so that the proof for Theorem(5.1.4) parallels exactly to that for Theorem(5.1.4)′.

Remark 5.1.5. We can also reduce Theorem(5.1.4) to Theorem(5.1.4)′ directly. In fact, according to the direct decomposition $\mathfrak{t} = \mathfrak{t}'' + \mathfrak{t}'$, we write $\lambda = \lambda'' + \lambda'$, and

$\rho(\mathfrak{u}) = \rho'' + \rho'$. If $\lambda \in \mathcal{F}$, then $\rho' = \rho(\mathfrak{u}')$ in this quarternionic case. Assume that

$$\langle \lambda, \alpha_{|\mathfrak{t}''} \rangle > 0 \qquad \text{for any } \alpha \in \Delta(\mathfrak{u}) \text{ with } \alpha_{|\mathfrak{t}''} \neq 0,$$

$$\lambda_i \geq -Q \quad \text{for any } i \ (1 \leq i \leq r+s).$$

(This condition is clearly weaker than $\lambda \in \mathcal{F}$.) Then we have

$$\begin{aligned} \left(\mathcal{R}_{\mathfrak{q}}^{\mathfrak{g}}\right)^{j+S-S'}(\mathbb{C}_\lambda) \neq 0 &\iff \left(\mathcal{R}_{\mathfrak{q}}^{\mathfrak{g}}\right)^{j+S-S'}(\mathbb{C}_{\lambda+\xi\rho''}) \neq 0 \quad (\forall \xi \in \mathbb{N}) \\ &\iff \left(\mathcal{R}_{\mathfrak{q}'}^{\mathfrak{g}'}\right)^{j}(\mathbb{C}_{\lambda'}) \neq 0 \end{aligned}$$

The first equivalence is derived from Lemma(3.4.1)(2). Indeed, we can easily find a sequence $\lambda^{(0)} = \lambda, \lambda^{(1)}, \ldots, \lambda^{(n)} = \lambda + \xi\rho''$ so that

$$\mathcal{A}\left(\lambda^{(i-1)} \triangleright \lambda^{(i)}\right) = \{\lambda^{(i)}\},\ \mathcal{A}\left(\lambda^{(i)} \triangleright \lambda^{(i-1)}\right) = \{\lambda^{(i-1)}\} \quad \text{for } i = 1, 2, \ldots, n.$$

The second equivalence is from Lemma(3.2.1)(2) and Fact(1.4.1)(2). Hence we have, in particular, $\left(\mathcal{R}_{\mathfrak{q}}^{\mathfrak{g}}\right)^{S}(\mathbb{C}_\lambda) \neq 0 \Leftrightarrow \left(\mathcal{R}_{\mathfrak{q}'}^{\mathfrak{g}'}\right)^{S'}(\mathbb{C}_{\lambda'}) \neq 0$. Since $Q' = Q$, we have $\lambda' \in \mathcal{B}$ and we are done.

Remark 5.1.6. Let $\tilde{\lambda} \equiv \tilde{\lambda}(\mathfrak{q}) := -\rho(\mathfrak{u}) + 2\rho(\mathfrak{u} \cap \mathfrak{k})$. We write the coordinates of $\tilde{\lambda}$ as $\tilde{\lambda}_i\,(1 \leq i \leq r+s)$ by restricting it to $\mathfrak{t}$. Then elementary arithmetic shows the equivalence of the following four conditions:

(5.1.7)(a) $$\tilde{\lambda} \in \mathcal{B}.$$

(5.1.7)(b) $$Q \geq \tilde{\lambda}_i > -Q \quad \text{for all } i \ (1 \leq i \leq r+s).$$

(5.1.7)(c) $$Q \geq \tilde{\lambda}_i \quad \text{for all } i \ (1 \leq i \leq r+s).$$

(5.1.7)(d) $$\begin{cases} \displaystyle\sum_{k=1}^{j} m_k \leq 2(q-s) + \sum_{k=1}^{j-1} n_k \quad (1 \leq j \leq M), \\ \displaystyle\sum_{k=1}^{j} n_k \leq 2(p-r) + \sum_{k=1}^{j} m_k \quad (1 \leq j \leq N). \end{cases}$$

The implication (5.1.7)(c) $\Rightarrow$ (5.1.7)(b) looks strange, but it turns out that the estimate on m_j and n_j blocks (see after (5.1.2) for notation) plays a 'complementary' role to each other with reverse signs.

5.2. key lemmas

Definition 5.2.1. For each subset S of $\mathcal{B}_{\mathbb{Z}}$, we define $\widetilde{S} \subset \mathcal{B}_{\mathbb{Z}}$ to be the smallest set such that

(5.2.2)(a) $$\widetilde{S} \supset S$$

(5.2.2)(b) $$\lambda \in \widetilde{S},\ \lambda' \in \mathfrak{t}^*_{\mathbb{Z}} \text{ and } \mathcal{A}(\lambda \triangleright \lambda') \subset \widetilde{S} \cup \{\lambda'\} \Rightarrow \lambda' \in \widetilde{S}.$$

By definition $\mathcal{A}(\lambda \triangleright \lambda')$ is a subset of $\mathfrak{h}^*$. However, as is easy to see, if λ, $\lambda' \in \mathcal{B}_{\mathbb{Z}}(\subset \mathfrak{t}^*)$, then $\mathcal{A}(\lambda \triangleright \lambda') \subset \mathcal{B}_{\mathbb{Z}}$, where we regard $\mathfrak{t}^*$ as a subspace of $\mathfrak{h}^*$ as usual. Finally, set

$$\mathcal{N} \equiv \mathcal{N}(\mathfrak{q}) := \{\lambda \in \mathcal{B}_{\mathbb{Z}}\,;\ \mathcal{R}^i_{\mathfrak{q}}(\mathbb{C}_\lambda) = 0 \text{ for all } i\}.$$

Here are two key lemmas.

Lemma 5.2.3. *With notation as above,*

1) $\widetilde{\widetilde{S}} = \widetilde{S}$ *for any subset S of $\mathcal{B}_{\mathbb{Z}}$.*

2) $\widetilde{\mathcal{N}} = \mathcal{N}$.

Lemma 5.2.4. $\widetilde{\{\lambda\}} = \mathcal{B}_{\mathbb{Z}}$ *if* $\lambda \in \mathcal{B} \cap \mathcal{F}_{\mathbb{Z}}$.

Proof of Lemma(5.2.3). Part (1) is clear from definition and Part (2) is a direct consequence of the relation between translation functor and cohomologically parabolic induction in Lemma(3.4.1)(1). $\square$

Admitting Lemma(5.2.4) for a while, we shall complete the proof of Theorem(5.1.4)$'$.

Proof of Theorem(5.1.4)'. If $\tilde\lambda \equiv -\rho(\mathfrak{u}) + 2\rho(\mathfrak{u}\cap\mathfrak{k}) \in \mathcal{B}$, then $\tilde\lambda \notin \mathcal{N}$ because

$$\sum_i (-1)^i \dim \operatorname{Hom}_K(\mathbf{1}, \mathcal{R}_{\mathfrak{q}}^{S-i}(\mathbb{C}_{\tilde\lambda})) = 1,$$

as is easy to see by the generalized Blattner formula. Thus $\mathcal{N} \subsetneqq \mathcal{B}_{\mathbb{Z}}$, which implies $\mathcal{B} \cap \mathcal{F}_{\mathbb{Z}} \subset \mathcal{F}_{\mathbb{Z}} \setminus \mathcal{N}$ because of Lemma(5.2.3) and Lemma(5.2.4). On the other hand, if $\lambda \in \mathcal{F}_{\mathbb{Z}}$ then $\mathbb{C}_\lambda$ is in the fair range and $\mathcal{R}_{\mathfrak{q}}^i(\mathbb{C}_\lambda) = 0$ for all $i \neq S$. Therefore $\mathcal{R}_{\mathfrak{q}}^S(\mathbb{C}_\lambda) \neq 0$ for any $\lambda \in \mathcal{B} \cap \mathcal{F}_{\mathbb{Z}}$. □

5.3. proof of the combinatorial part

We are now left with the proof of Lemma(5.2.4), whose statement depends only on $r+s$ and Q, and dose not depend on the particular partitions of r and s. Set $n := r+s$. Let $\mathfrak{t}_{\mathbb{R}}^* := \{\lambda = (\lambda_1, \dots, \lambda_n) \in \mathfrak{t}^* \,;\, \lambda_i \in \mathbb{R}\}$. Each connected component of

$$\{\lambda = (\lambda_1, \dots, \lambda_n) \in \mathfrak{t}_{\mathbb{R}}^* \,;\, \lambda_i \neq \lambda_j \,(i \neq j),\, \lambda_i \neq 0\} \subset \mathfrak{t}_{\mathbb{R}}^*$$

is a Weyl chamber for $W(C_n)$.

Claim(5.3.1). *Let $C \subset \mathfrak{t}_{\mathbb{R}}^*$ be a Weyl chamber. If $\lambda \in \mathcal{B}_{\mathbb{Z}} \cap C$, then $\widetilde{\{\lambda\}} \supset \mathcal{B}_{\mathbb{Z}} \cap \overline{C}$.*

Proof. Fix $\lambda' \in \mathcal{B}_{\mathbb{Z}} \cap C$. Then we can easily find a sequence $\lambda = \lambda^{(0)}, \lambda^{(1)}, \dots, \lambda^{(k)} = \lambda'$ such that $\mathcal{A}\left(\lambda^{(i-1)} \triangleright \lambda^{(i)}\right) = \{\lambda^{(i)}\}$ $(1 \leq i \leq k)$. Now the property(5.2.2) shows $\lambda' \in \widetilde{\{\lambda\}}$. □

Claim(5.3.2). *Let $\lambda = (n, n-1, \dots, 1)$. Then $\widetilde{\{\lambda\}}$ contains λ' in each case below.*

a) $\lambda' = (n, n-1, \dots, j, j+1, \dots, 1)$ $(1 \leq j \leq n-1)$

b) $\lambda' = (n, n-1, \dots, 2, -1)$

Proof. $\mathcal{A}(\lambda \triangleright \lambda'') = \{\lambda''\}$ and $\mathcal{A}(\lambda'' \triangleright \lambda') = \{\lambda, \lambda'\}$ if we choose λ'' as follows:

a) $\lambda'' := (n, n-1, \dots, j, j, \dots, 1)$,

b) $\lambda'' = (n, n-1, \ldots, 2, 0)$.

Now the claim follows from the property(5.2.2). $\square$

Claim(5.3.1) and Claim(5.3.2) lead us to

$$\widetilde{\{\lambda\}} = \mathcal{B}_{\mathbb{Z}} \tag{5.3.3}$$

for each fixed $\lambda \in \mathcal{B}_{\mathbb{Z}} \cap \mathcal{C}$, and for each Weyl chamber $\mathcal{C}$. Now Lemma(5.2.4) is proved.

§6. Proof of irreducibility results

6. irreducibility in the fair range

In this section, we prove a result about irreducibility of certain series of cohomologically induced representations. Suppose we are in the setting of §1.1-3. One might expect that

(6.1) If a metapletic representation $\mathbb{C}_\lambda$ of $(\mathfrak{l}, (L \cap K)^\sim)$ is in the weakly fair range, then the derived functor module $\mathcal{R}_{\mathfrak{q}}^S(\mathbb{C}_\lambda)$ is irreducible as a $(\mathfrak{g}, K)$-module or zero.

This is known to be false in general as alluded to in Fact(1.4.2)(2). However, it still holds in some interesting cases:

0) $\mathbb{C}_\lambda$ is in the weakly good range (see Fact(6.2.4)(c)).

i) $\mathfrak{g}$ is of type A (see Fact(6.2.4)(b)).

ii) $(\mathfrak{g}_0, \mathfrak{q})$ arises in a natural description of discrete series for a semisimple symmetric space ([33], see Fact(1.5.2) for notation).

In this section we shall prove a positive answer for (6.1) when $\mathbb{C}_\lambda$ is in the fair range and when

iii) $\mathfrak{g}$ is of type C and $[\mathfrak{l}, \mathfrak{l}]$ is also of type C.

To prove this, we need a slight refinement (Theorem(6.3.1)) of [33] Theorem 5.11, which assures a nice behavior under translation in an extremely special direction when applied to some Harish-Chandra bimodules: namely, twisted differential operators coming from $U(\mathfrak{g})$ when $\mathfrak{g} = \mathfrak{sp}(n,\mathbb{C})$.

Our approach parallels exactly Vogan's proof there. But we isolate the necessary steps explicitly since it is complicated in detail (ex. (6.7.7)).

6.2. twisted differential operators

We introduce some notation used throughout this section §6. For materials of this subsection, we refer to [2], [3], [6], [30] and [33].

Fix a connected complex reductive (algebraic) group $G_{\mathbb{C}}$ with Lie algebra $\mathfrak{g}$. We identify Harish-Chandra bimodules with actual Harish-Chandra modules for $G_{\mathbb{C}}$ (regarded as a real group) by a Chevalley anti-automorphism. Fix a Cartan subalgebra $\mathfrak{h}$ and a parabolic subalgebra $\mathfrak{q}$ of $\mathfrak{g}$ with Levi decomposition $\mathfrak{q} = \mathfrak{l} + \mathfrak{u}$ so that $\mathfrak{l}$ contains $\mathfrak{h}$. Fix a weakly fair weight $\gamma \in \mathfrak{h}^*$. Write $R_\gamma(\mathfrak{l} : \mathfrak{l})$ for the one-dimensional Harish-Chandra bimodule for $L_{\mathbb{C}}$ with $\mathfrak{Z}(\mathfrak{l} \otimes \mathbb{C})$-infinitesimal character $(\gamma,\gamma) \in \mathfrak{h}^* \oplus \mathfrak{h}^*$. That is, its bimodule structure is given by,

$$X \cdot y = y \cdot X = \gamma(X_1)y$$

for $y \in R_\gamma(\mathfrak{l} : \mathfrak{l})$, $X = X_1 + X_2 \in \mathfrak{t} \oplus [\mathfrak{l},\mathfrak{l}] = \mathfrak{l}$ (notation (1.1.2)). Define

$$R_\gamma(\mathfrak{l} : \mathfrak{g}) = \operatorname{Ind}(Q_{\mathbb{C}} \uparrow G_{\mathbb{C}})(R_\gamma(\mathfrak{l} : \mathfrak{l})), \tag{6.2.1}$$

a Harish-Chandra bimodule for $G_{\mathbb{C}}$, which has a unique $G_{\mathbb{C}}$-fixed vector. Here we denote by $\operatorname{Ind}(Q_{\mathbb{C}} \uparrow G_{\mathbb{C}})$ normalized parabolic induction for Harish-Chandra bimodules.

Note that $R_\gamma(\mathfrak{l} : \mathfrak{g})$ may be identified with a $G_{\mathbb{C}}$-equivariant twisted ring of differential operators on the generalized flag variety $G_{\mathbb{C}}/Q_{\mathbb{C}}$. Then $R_\gamma(\mathfrak{l} : \mathfrak{g})$ has a natural algebra structure with

$$d : U(\mathfrak{g}) \to R_\gamma(\mathfrak{l} : \mathfrak{g}) \tag{6.2.2}$$

an algebra homomorphism that is nonzero on the unique $G_{\mathbb{C}}$-fixed vector. Define

$$\text{(6.2.3)(a)} \qquad I_\gamma(\mathfrak{l}:\mathfrak{g}) := \operatorname{Ker}(d : U(\mathfrak{g}) \to R_\gamma(\mathfrak{l}:\mathfrak{g}))$$

$$\text{(6.2.3)(b)} \qquad A_\gamma(\mathfrak{l}:\mathfrak{g}) := \operatorname{Image}(d : U(\mathfrak{g}) \to R_\gamma(\mathfrak{l}:\mathfrak{g})) \simeq U(\mathfrak{g})/I_\gamma(\mathfrak{l}:\mathfrak{g}).$$

If $\mathfrak{q}$ is a Borel subalgebra, we write simply $I(\gamma)$ for $I_\gamma(\mathfrak{h}:\mathfrak{g})$, which is nothing but the ideal in $U(\mathfrak{g})$ generated by the corresponding maximal ideal in $\mathfrak{Z}(\mathfrak{g})$. From this view point, we put $I(w\cdot\gamma) = I(\gamma)$ $(w \in W(\mathfrak{g},\mathfrak{h}))$ so that $I(\gamma)$ is defined for all $\gamma \in \mathfrak{h}^*$. Here is the irreducibility result due to J.Bernstein that we need (see [30] Proposition 16.8, [2] Proposition III.2.2.2 and I.5.6, see also [33] Proposition 5.7).

Fact 6.2.4. *Retain notations as above.*

a) *Suppose that W is a one dimensional metapletic $(\mathfrak{l}, (L\cap K)^{\sim})$ representation having $\mathfrak{Z}(\mathfrak{l})$-infinitesimal character γ in the weakly fair range. The algebra $R_\gamma(\mathfrak{l}:\mathfrak{g})$ acts on $\mathcal{R}_{\mathfrak{q}}^S(W)$, and the resulting $(R_\gamma(\mathfrak{l}:\mathfrak{g}), K)$-module is irreducible or zero.*

b) *If the moment map*

$$\pi : T^*(G_{\mathbb{C}}/Q_{\mathbb{C}}) \simeq G_{\mathbb{C}} \underset{Q_{\mathbb{C}}}{\times} \mathfrak{q}^{\perp} \to G\cdot(\mathfrak{q}^{\perp}) \subset \mathfrak{g}^*$$

is birational and has a normal image, then $A_\gamma(\mathfrak{l}:\mathfrak{g}) = R_\gamma(\mathfrak{l}:\mathfrak{g})$.

c) *If γ is in the weakly good range, then $A_\gamma(\mathfrak{l}:\mathfrak{g}) = R_\gamma(\mathfrak{l}:\mathfrak{g})$.*

If $A_\gamma(\mathfrak{l}:\mathfrak{g}) = R_\gamma(\mathfrak{l}:\mathfrak{g})$, then Part (a) implies the irreducibility (or vanishing) of $\mathcal{R}_{\mathfrak{q}}^S(\mathbb{C}_\lambda)$ as a $(\mathfrak{g}, K)$-module. It is known that the assumption in Fact(6.2.4)(b) is satisfied if $\mathfrak{g} = \mathfrak{gl}(n,\mathbb{C})$ and $\mathfrak{q}$ is any parabolic (see [17]) or if $\mathfrak{g} = \mathfrak{so}(n,\mathbb{C})$ and $\mathfrak{q}$ is of the form in §2.5 (see [12] for birationality and [18] for normality, see also [2] III.3.2), whence Part (5) of Theorem 2, 3. But unfortunately, the birationality fails (although the image of π is normal) when $\mathfrak{g} = \mathfrak{sp}(n,\mathbb{C})$ and $\mathfrak{q}$ is of the form in §2.1.

This should be compared with the fact that the normality frequently fails (although π is birational) in the setting of Fact(1.5.2) (discrete series for semisimple symmetric spaces) if $\mathfrak{g} = \mathfrak{sp}(n,\mathbb{C})$. Thus, throughout the rest of this section we shall restrict ourselves to the case where $\mathfrak{g} = \mathfrak{sp}(n,\mathbb{C})$ and $\mathfrak{q}$ is a general parabolic subalgebra and to the case of singular $\mathfrak{Z}(\mathfrak{g})$-infinitesimal characters.

6.3. theorem

Theorem 6.3.1 (cf. [33] **Theorem 5.11).** *Suppose $\mathfrak{g}$ is $\mathfrak{sp}(n,\mathbb{C})$. Let $\mathfrak{h}$ be the standard Cartan subalgebra identified with $\mathbb{C}^n$ as usual. Fix non-negative integers N, r such that $1 \le N \le n-r$. Write $l = n - N - r$. Let $\mathfrak{q}$ be the standard parabolic subalgebra with Levi factor*

$$\mathfrak{l} = \mathfrak{gl}(N,\mathbb{C}) \oplus \mathfrak{gl}(1,\mathbb{C})^r \oplus \mathfrak{sp}(l,\mathbb{C}).$$

Suppose γ is the $\mathfrak{Z}(\mathfrak{l})$-infinitesimal character of a one-dimensional representation of $\mathfrak{l}$:

(6.3.2)(a) $$\gamma = (\lambda_0 + N - 1, \lambda_0 + N - 2, \dots, \lambda_0, \lambda_1, \dots, \lambda_r, l, l-1, \dots, 1).$$

Assume that

(6.3.2)(b) $$\lambda_j \in \mathbb{Z} \text{ for any } j, \quad l \ge \lambda_0 \ge \cdots \ge \lambda_r \ge 0 \quad \text{and} \quad \lambda_0 > 0.$$

Set $\xi := (l-\lambda_0)^N \oplus 0^{n-N} = (l-\lambda_0, \dots, l-\lambda_0, 0, \dots, 0)$. Suppose M is an irreducible $\mathfrak{g}$-module annihilated by $I_{\gamma+\xi}(\mathfrak{l} : \mathfrak{g})$. Then $\psi^{\gamma}_{\gamma+\xi} M$ is an irreducible $\mathfrak{g}$-module or zero.

Remark 6.3.3. Theorem 5.11 in [33] is the case where $N = 2$. But the above theorem with $N = 1$ may also apply to the irreducibility results in [33] owing to the Borel-Weil-Bott theorem for $U(2)$. In order to prove Part(5) of Theorem 1 (§2.2), we need the case where $N = 1$.

6.4. irreducibility result

Corollary 6.4.1. *Let G be a real form of $Sp(n,\mathbb{C})$, K a maximal compact subgroup of G and θ the corresponding Cartan involution of G. Let $\mathfrak{p} = \mathfrak{m} + \mathfrak{n}$ be a θ-stable parabolic subalgebra (see* §1.1) *of $\mathfrak{g}$ with a Levi factor*

$$\mathfrak{m} \simeq \mathfrak{gl}(n_1,\mathbb{C}) \oplus \mathfrak{gl}(n_2,\mathbb{C}) \oplus \cdots \oplus \mathfrak{gl}(n_k,\mathbb{C}) \oplus \mathfrak{sp}(l,\mathbb{C}),$$

for some $0 \le l \le n$ and $n_j \in \mathbb{N}_+$ such that $\sum_{j=1}^{k} n_j = n - l$. Fix a Cartan subalgebra $\mathfrak{h}$ of $\mathfrak{m}$. Let W be a metapletic $(\mathfrak{m}, (M \cap K)^{\sim})$-character with $\mathfrak{Z}(\mathfrak{m})$-infinitesimal character $\gamma = (\gamma_1, \ldots, \gamma_{n-l}, l, l-1, \ldots, 1) \in \mathfrak{h}^*$. Assume that $\gamma_j \in \mathbb{Z}$ and

$$\gamma_1 \ge \gamma_2 \ge \cdots \ge \gamma_{n-l} > 0.$$

Then $\mathcal{R}_{\mathfrak{p}}^{s}(W)$ is an irreducible $(\mathfrak{g}, K)$-module or zero.

This includes the irreducibility assertions in Part (5) of Theorem 1. In fact, it is obtained by applying the above corollary to the case when $n_1 = \cdots = n_k = 1$.

6.5. Vogan's idea on the translation principle for $A_\gamma(\mathfrak{l} : \mathfrak{g})$

We shall review shortly a technique due to Vogan [33] which shows irreducibility of certain series of derived functor modules. The main tools for reduction to good parameters are a combination of induction by stages and translation principle. Translation is used in a pair of translation functors for Harish-Chandra modules and Harish-Chandra bimodules.

Let R be a complex algebra, endowed with Harish-Chandra bimodule structure for $G_{\mathbb{C}}$ through an algebra homomorphism

$$d : U(\mathfrak{g}) \to R. \tag{6.5.1}$$

Let M be an R-module which is $\mathfrak{Z}(\mathfrak{g})$-finite via (6.5.1). Let F be a finite dimensional representation of $\mathfrak{g}$ with extremal weight $-\xi$ on which the adjoint action of $\mathfrak{g}$ exponentiates to $G_{\mathbb{C}}$. A formal argument (the Jacobson density theorem) (see [33] Corollary

3.9) shows that

(6.5.2) If M is an irreducible R-module, then $\psi^{\gamma}_{\gamma+\xi}M$ is an irreducible $\left(\psi^{(\gamma,\gamma)}_{(\gamma+\xi,\gamma+\xi)}R\right)$-module or zero.

Retain notations in §6.2. Let $\gamma := \lambda + \rho_{\mathfrak{l}} \in \mathfrak{h}^*$ be in the weakly fair range. Let us apply (6.5.2) to $d : U(\mathfrak{g}) \twoheadrightarrow A_{\lambda+\rho_{\mathfrak{l}}}(\mathfrak{l} : \mathfrak{g})$. We expect a reasonable behavior under translation (see Theorem(6.3.1)):

$$\psi^{(\lambda+\rho_{\mathfrak{l}},\lambda+\rho_{\mathfrak{l}})}_{(\lambda+\rho_{\mathfrak{l}}+\xi,\lambda+\rho_{\mathfrak{l}}+\xi)}A_{\lambda+\rho_{\mathfrak{l}}+\xi}(\mathfrak{l} : \mathfrak{g}) = A_{\lambda+\rho_{\mathfrak{l}}}(\mathfrak{l} : \mathfrak{g}). \tag{6.5.3}$$

But, in the (weakly) fair range, the behavior of $A_{\lambda+\rho_{\mathfrak{l}}}(\mathfrak{l} : \mathfrak{g})$ under translation is not as good as those of $\mathcal{R}^S_{\mathfrak{q}}(\mathbb{C}_\lambda)$ and $R_{\lambda+\rho_{\mathfrak{l}}}(\mathfrak{l} : \mathfrak{g})$. Vogan pointed out that (6.5.3) is guaranteed by the existence of Harish-Chandra bimodules C_λ and $C_{\lambda+\xi}$ satisfying

(6.5.4)(a) $\psi^{(\lambda+\rho_{\mathfrak{l}},\lambda+\rho_{\mathfrak{l}})}_{(\lambda+\rho_{\mathfrak{l}}+\xi,\lambda+\rho_{\mathfrak{l}}+\xi)}C_{\lambda+\xi} = C_\lambda$.

(6.5.4)(b) Each of $C_{\lambda+\xi}$ and C_λ is generated by its unique $G_{\mathbb{C}}$-fixed vector as a Harish-Chandra bimodule.

(6.5.4)(c) There are Harish-Chandra bimodule maps

$$C_{\lambda+\xi} \to R_{\lambda+\rho_{\mathfrak{l}}+\xi}(\mathfrak{l} : \mathfrak{g})$$

$$C_\lambda \to R_{\lambda+\rho_{\mathfrak{l}}}(\mathfrak{l} : \mathfrak{g})$$

that are non-zero on the $G_{\mathbb{C}}$-fixed vectors.

With a special choice of ξ and C_λ in §6.7, we shall check these conditions in §6.8-10.

6.6. notations about $GL(n,\mathbb{C})$ and $Sp(n,\mathbb{C})$

Since we shall treat only complex reductive Lie groups in §6.6-10, we write $GL(n)$ for $GL(n,\mathbb{C})$ and etc, for simplicity.

First let $G_{\mathbb{C}} = GL(n)$. Take a Cartan subalgebra $\mathfrak{h}$ of $\mathfrak{g}$ so that $\mathfrak{h}$ consists of $n \times n$ diagonal matrices with complex entries. Diagonal coordinates induce $\mathfrak{h}^* \simeq \mathbb{C}^n$ as usual. Let B be a Borel subgroup of $G_{\mathbb{C}}$ consisting of upper triangular matrices. Let $\gamma = (\gamma_1, \dots, \gamma_n) \in \mathfrak{h}^*$. Define a Harish-Chandra bimodule for $G_{\mathbb{C}} = GL(n)$ by,

$$[\gamma] \equiv [\gamma_1, \dots, \gamma_n] = \text{unique irreducible subquotient of } R_\gamma(\mathfrak{h} : \mathfrak{g})$$
$$\text{containing the } G_{\mathbb{C}}\text{-fixed vectors.}$$

We remark that $[\gamma] = J(2\gamma)$ in the notation of [30] §11 when regarded as an irreducible spherical representation of $G_{\mathbb{C}}$ (as a real group).

Assume that the sequence $(\operatorname{Re} \gamma_j)$ is decreasing. With the above parametrization, we have (see [30] Theorem 11.5, Lemma 11.11):

(6.6.1)(a) $[\gamma]$ is the unique irreducible quotient of $R_\gamma(\mathfrak{h} : \mathfrak{g})$.

(6.6.1)(b) $[\gamma]$ is finite dimensional iff $\gamma_j - \gamma_{j+1} \in \mathbb{N}_+$ for any j.

(6.6.1)(c) $[\gamma]$ is one dimensional iff $\gamma = [c; -1, n]$ (§3.5) for some $c \in \mathbb{C}$.

Next we look upon $GL(n)$ as a reductive subgroup in $Sp(n)$. Since $GL(n)$ is of maximal rank in $Sp(n)$, $\mathfrak{h}$ is also a Cartan subalgebra in $\mathfrak{sp}(n)$. Let $\tilde{B}(\supset B)$ be the Borel subgroup of $Sp(n)$, making $\langle n \rangle = (n, n-1, \dots, 1)$ dominant for $\Delta(\tilde{\mathfrak{b}}, \mathfrak{h})$.

Definition 6.6.2. An (ordered) partition of n is a sequence

$$\pi = (p_1, \dots, p_s)$$

of non-negative integers, such that $\sum_{j=1}^{s} p_j = n$. Denote by $P(\pi)$ the parabolic subgroup of $GL(n)$ containing B and having a Levi factor

$$L(\pi) \stackrel{\text{def}}{=} GL(p_1) \times GL(p_2) \times \cdots \times GL(p_s). \tag{6.6.3}$$

Similarly we denote by $\tilde{P}(\pi)$ the parabolic subgroup of $Sp(n)$ containing $\tilde{B}$ with the same Levi factor $L(\pi)$.

6.7. definition of C_λ

Suppose that we are given $r,\, l \in \mathbb{N}_+,\ \lambda_j \in \mathbb{Z}\ (0 \leq j \leq r)$ such that,

$$l \geq \lambda_0 \geq \lambda_1 \geq \cdots \geq \lambda_r \geq 0, \tag{6.7.1(a)}$$

$$\lambda_0 > 0. \tag{6.7.1(b)}$$

Let $\{\Lambda_j\}_{1\leq j\leq s+1}$ be the totality of distinct values in $\{\lambda_j\,;\, 0 \leq j \leq r\} \cup \{0\}$ such that

$$\Lambda_1 > \cdots > \Lambda_s > \Lambda_{s+1} = 0. \tag{6.7.2(a)}$$

We shall ignore Λ_{s+1} unless $\lambda_r = 0$. The next definition serves to simplify the notation somewhat:

$$\Lambda_0 := l + 1. \tag{6.7.2(b)}$$

Set

$$m(j) := \sharp\{k\,;\, \lambda_k = \Lambda_j,\, 1 \leq k \leq r\} \in \mathbb{N} \qquad (1 \leq j \leq s+1). \tag{6.7.3}$$

Clearly $\sum_{j=1}^{s+1} m(j) = r \geq s-1$ (or $\geq s$ when $\lambda_r = 0$). Let

$$\lambda^{(0)} = [\lambda_0 + N - 1; -1, N] := (\lambda_0 + N - 1, \lambda_0 + N - 2, \ldots, \lambda_0).$$

Define

$$\begin{aligned} \nu^{(j)} :&= [\Lambda_{j-1} - 1; -1, \Lambda_{j-1} - \Lambda_j] \\ &= (\Lambda_{j-1} - 1, \Lambda_{j-1} - 2, \ldots, \Lambda_j) \quad (1 \leq j \leq s). \end{aligned} \tag{6.7.4(a)}$$

If $\Lambda_s \geq 2$, we also define

$$\begin{aligned} \nu^{(s+1)} &:= [\Lambda_s - 1; -1, \Lambda_s - 1] \equiv \langle \Lambda_s - 1 \rangle \\ &= (\Lambda_s - 1, \ldots, 2, 1). \end{aligned} \tag{6.7.4(b)}$$

With this notation, if $\{\lambda_1,\dots,\lambda_r\} \ni 1$ (equivalently, if $\Lambda_s = 1$), then $(\nu^{(1)},\nu^{(2)},\dots,\nu^{(s)})$ is a grouping of $(l, l-1,\dots,2,1)$. If $\{\lambda_1,\dots,\lambda_r\} \not\ni 1$ (equivalently, if $\Lambda_s \geq 2$), then $(\nu^{(1)},\nu^{(2)},\dots,\nu^{(s)},\nu^{(s+1)})$ is a grouping of $(l, l-1,\dots,2,1)$. If $\{\lambda_1,\dots,\lambda_r\} \ni 0$ (equivalently, if $\lambda_r = \Lambda_{s+1} = 0$ and $m(s+1) > 0$), then $({\Lambda_1}^{m(1)}, {\Lambda_2}^{m(2)},\dots,{\Lambda_s}^{m(s)}, 0^{m(s+1)})$ is a grouping of $(\lambda_1,\lambda_2,\dots,\lambda_r)$. If $\{\lambda_1,\dots,\lambda_r\} \not\ni 0$ (equivalently, if $\lambda_r \neq 0$), then $({\Lambda_1}^{m(1)}, {\Lambda_2}^{m(2)},\dots,{\Lambda_s}^{m(s)})$ is a grouping of $(\lambda_1,\lambda_2,\dots,\lambda_r)$.

Put

$$\pi := (N, 1^{m(1)}, \Lambda_0 - \Lambda_1, \dots, 1^{m(s)}, \Lambda_{s-1} - \Lambda_s, \Lambda_s - 1, 1^{m(s+1)}), \tag{6.7.5}$$

an ordered partition of $n = N + r + l$. Then according to notation (6.6.3), a reductive subgroup $L(\pi)$ of $Sp(n)$ is given by,

$$GL(N) \times \prod_{j=1}^{s} \left(GL(1)^{m(j)} \times GL(\Lambda_{j-1} - \Lambda_j) \right) \times GL(\Lambda_s - 1) \times GL(1)^{m(s+1)}.$$

More precisely, if $\{\lambda_1,\dots,\lambda_r\} \ni 1$ (equivalently, if $\Lambda_s = 1$), then the above $GL(\Lambda_s - 1)$ and the below $[\nu^{(s+1)}]$ should be omitted. Similarly, if $\lambda_r > 0$ (equivalently, if $m(s+1) = 0$), then the above $GL(1)^{m(s+1)}$ and the below $[\Lambda_{s+1}]^{m(s+1)}$ should be omitted. We define a Harish-Chandra bimodule for $L(\pi)$ by

$$\sigma_\lambda \stackrel{\text{def}}{=} [\lambda^{(0)}] \otimes \bigotimes_{j=1}^{s} \left([\Lambda_j]^{m(j)} \otimes [\nu^{(j)}] \right) \otimes [\nu^{(s+1)}] \otimes [\Lambda_{s+1}]^{m(s+1)}, \tag{6.7.6}$$

and a Harish-Chandra bimodule for $Sp(n)$ by

$$C_\lambda \stackrel{\text{def}}{=} \operatorname{Ind}(\tilde{P}(\pi) \uparrow Sp(n))(\sigma_\lambda). \tag{6.7.7}$$

Finally, take

$$\xi := (l - \lambda_0)^N \oplus 0^{n-N}. \tag{6.7.8}$$

We define $\sigma_{\lambda+\xi}$ and $C_{\lambda+\xi}$ similarly by changing only $\lambda^{(0)}$. We shall prove Theorem(6.3.1) by checking (6.5.4)(a)-(c) for these C_λ, $C_{\lambda+\xi}$ in the subsequent three subsections.

The somewhat complicated notation of Λ_i, $m(i)$, $\nu^{(j)}$ and π will be used throughout §6. The following figure would be helpful useful to memorize the definition.

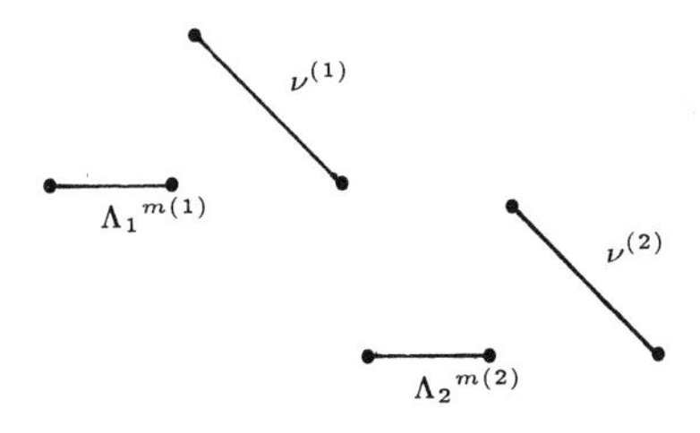

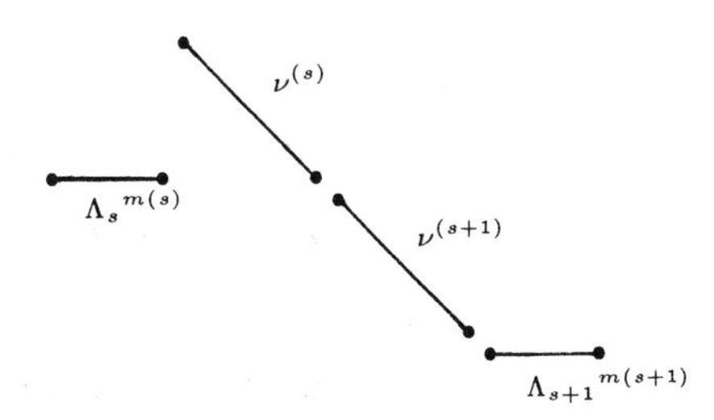

Figure 6.7.9

6.8. verification of (6.5.4)(a)

The proof of (6.5.4)(a) depends on the following

Lemma 6.8.1. *In the setting of* (6.7.1)-(6.7.5),(6.7.8), *set*

$$\gamma := (\lambda^{(0)}, \Lambda_1{}^{m(1)}, \nu^{(1)}, \Lambda_2{}^{m(2)}, \nu^{(2)}, \ldots, \Lambda_s{}^{m(s)}, \nu^{(s)}, \nu^{(s+1)}, 0^{m(s+1)}).$$

Let $\mu \in \mathfrak{h}^* \simeq \mathbb{C}^n$ *be a weight in* $F(Sp(n), -\xi)$. *Assume that*

$$\begin{cases} \mu \text{ is dominant for } \Delta^+(\mathfrak{l}(\pi), \mathfrak{h}), \\ \gamma + \xi + \mu = w \cdot \gamma \quad \text{for some } w \in W(\mathfrak{sp}(n)). \end{cases}$$

Then

$$\mu = -\xi.$$

Write $\rho_{\mathfrak{l}(\pi)} = \rho(\Delta^+(\mathfrak{l}(\pi)))$. We note that the above conclusion is equivalent to

$$\mathcal{A}(\Delta^+(\mathfrak{l}(\pi)); \gamma + \xi - \rho_{\mathfrak{l}(\pi)} \triangleright \gamma - \rho_{\mathfrak{l}(\pi)}) = \{\gamma - \rho_{\mathfrak{l}(\pi)}\}$$

in terms of our definition in §3.3 for a pair $\mathfrak{l}(\pi) \subset \mathfrak{sp}(n)$.

Assuming this lemma for a while, we verify (6.5.4)(a).

Proof of (6.5.4)(a). Set $\gamma = \lambda + \rho_{\mathfrak{l}}$ as in §6.5.

$$\begin{aligned} &\psi^{(\gamma,\gamma)}_{(\gamma+\xi,\gamma+\xi)} C_{\lambda+\xi} \\ =&P_{(\gamma,\gamma)}\left(\operatorname{Ind}(\tilde{P}(\pi) \uparrow Sp(n))(\sigma_{\lambda+\xi}) \otimes \operatorname{End}(F(\mathfrak{g}, -\xi))\right) \\ \simeq&P_{(\gamma,\gamma)}\left(\operatorname{Ind}(\tilde{P}(\pi) \uparrow Sp(n))(\sigma_{\lambda+\xi} \otimes \operatorname{End}(F(\mathfrak{g}, -\xi)))\right). \end{aligned}$$

As the above Lemma(6.8.1) assures that

$$\begin{aligned} &\mathcal{A}\left((\gamma + \xi - \rho_{\mathfrak{l}(\pi)}, \gamma + \xi - \rho_{\mathfrak{l}(\pi)}) \triangleright (\gamma - \rho_{\mathfrak{l}(\pi)}, \gamma - \rho_{\mathfrak{l}(\pi)})\right) \\ =&\{(\gamma - \rho_{\mathfrak{l}(\pi)}, \gamma - \rho_{\mathfrak{l}(\pi)})\} \end{aligned}$$

for $\mathfrak{l}(\pi) \otimes_{\mathbb{R}} \mathbb{C} \subset \mathfrak{sp}(n, \mathbb{C}) \otimes_{\mathbb{R}} \mathbb{C}$, the last Harish-Chandra bimodule is isomorphic to $\operatorname{Ind}(\tilde{P}(\pi) \uparrow Sp(n))(\sigma_\lambda) = C_\lambda$ (cf. Lemma(3.4.1)(2)). Hence (6.5.4)(a). □

Lemma(6.8.1) is deduced from the following two elementary claims.

Claim 6.8.2. *Let $\gamma = (\gamma_1, \dots, \gamma_n)$, $\xi = L^N \oplus 0^{n-N} \in \mathbb{Z}^n$ satisfy $\gamma_k \geq 0$ for any k and $L \geq 0$. Let $\mu \in \mathbb{Z}^n$ be a weight of $F(Sp(n), -\xi)$. Assume that*

$$\gamma + \xi + \mu = w \cdot \gamma \tag{6.8.3}$$

for some w, an element of the Weyl group $W(\mathfrak{sp}(n)) \simeq \mathfrak{S}_n \ltimes {\mathbb{Z}_2}^n$. Then the following holds.

(6.8.4)(a) $\quad -L \leq \mu_k \leq 0 \quad$ *for any k $(1 \leq k \leq n)$.*

(6.8.4)(b) $\quad \sum_{k=1}^{n} \mu_k = -NL.$

(6.8.4)(c) $\quad$ (6.8.3) *holds for some $w \in \mathfrak{S}_n$.*

(6.8.4)(d) $\quad$ *If $N+1 \leq k \leq n$ and $\gamma_k \leq \min_{1 \leq j \leq N} \gamma_j$, then $\mu_k = 0$.*

Proof. As every weight of $F(Sp(n), -\xi)$ lies in the convex hull of the extremal weights, we have

$$-L\min(N, \#I) \leq \sum_{j \in I} \mu_j \leq L\min(N, \#I), \tag{6.8.5}$$

for any subset $I \subset \{1, \dots, n\}$. In particular, applying the above inequality to $I = \{1, \dots, n\}$, we have

$$\sum_{k=1}^{n} (\gamma + \xi + \mu)_k = \sum_{k=1}^{n} \gamma_k + NL + \sum_{k=1}^{n} \mu_k \geq \sum_{k=1}^{n} \gamma_k + NL - NL = \sum_{k=1}^{n} \gamma_k.$$

On the other hand, since $\gamma_k \geq 0$ for any k, we have

$$\sum_{k=1}^{n} (\gamma + \xi + \mu)_k = \sum_{k=1}^{n} (w \cdot \gamma)_k \leq \sum_{k=1}^{n} \gamma_k,$$

from (6.8.3). Thus the both sides must be equal, so that we have (6.8.4)(b),(c). Now (6.8.4)(a) follows from (6.8.4)(b) by applying (6.8.5) to $I = \{1, \dots, n\} \setminus \{k\}$ and $I = \{k\}$.

It remains to check (6.8.4)(d). From (6.8.4)(a), we have

$$(\gamma+\xi+\mu)_j = \begin{cases} \gamma_j + L + \mu_j \geq \gamma_j & \text{if } 1 \leq j \leq N, \\ \gamma_j + \mu_j \leq \gamma_j & \text{if } N+1 \leq j \leq n. \end{cases}$$

Fix k such that $\gamma_k \leq \min_{1\leq j\leq N} \gamma_j$. Then

$$\begin{aligned} &\{j\,;\, 1 \leq j \leq n,\ (\gamma+\xi+\mu)_j \geq \gamma_k\} \\ =&\{1,2,\dots,N\} \cup \{j\,;\, N+1 \leq j \leq n, (\gamma+\mu)_j \geq \gamma_k\} \\ \subset&\{1,2,\dots,N\} \cup \{j\,;\, N+1 \leq j \leq n, \gamma_j \geq \gamma_k\} \\ =&\{j\,;\, 1 \leq j \leq n,\ \gamma_j \geq \gamma_k\}. \end{aligned}$$

Since $\gamma+\xi+\mu$ is a permutation of γ ((6.8.4)(c)), both sets must coincide. Thus we have

$$\gamma_j \geq \gamma_k \quad \Rightarrow \quad \gamma_j + \mu_j \geq \gamma_k$$

for any j with $N+1 \leq j \leq n$. Therefore if we assume moreover $N+1 \leq k \leq n$, then we have $\gamma_k + \mu_k \geq \gamma_k$ by taking $j = k$. Hence $\mu_k = 0$ from (6.8.4)(a). □

Claim 6.8.6. *Suppose we are in the setting of* Claim(6.8.2) *with the same hypotheses. Set* $J := \{k\,;\, \gamma_k > \min_{1\leq j\leq N} \gamma_j\}$ *and* $J' := \{k\,;\, N+1 \leq k \leq n,\ k \notin J\}$. *Assume moreover that*

$$\min_{j\in J} \mu_j \geq \min_{j\in J'} \mu_j, \tag{6.8.7}$$

If J' is not empty, then $\mu = -\xi$.

Proof. First we show that $\mu_j = 0$ $(N+1 \leq j \leq n)$ if $J' \neq \emptyset$. Indeed, $\mu_j = 0$ when $j \in J'$ by (6.8.4)(d). If $J' \neq \emptyset$, this implies $\mu_j = 0$ $(j \in J)$ because of (6.8.4)(a) and (6.8.7). Therefore $\mu_j = 0$ for $j \in J \cup J' = \{N+1, N+2, \dots, n\}$. Now (6.8.4)(a),(b) determines $\mu_1 = -L$. □

6.9. verification of (6.5.4)(b)

Retain notations in §6.7. Set

$$n(j) := \begin{cases} N + m(1) + \Lambda_0 - \Lambda_1 & (j=1), \\ m(j) + \Lambda_{j-1} - \Lambda_j & (2 \le j \le s). \end{cases}$$

Proof of (6.5.4)(b). Define the partition $\tilde{\pi}$ of n by

$$\tilde{\pi} := (n(1), \ldots, n(s), \Lambda_s - 1, 1^{m(s+1)}),$$

and associate a parabolic subgroup $\tilde{P}(\tilde{\pi})$ of $Sp(n)$ (Definition(6.6.2)) with the Levi factor

$$L(\tilde{\pi}) = \left(\prod_{j=1}^{s} G_j\right) \times GL(\Lambda_s - 1) \times GL(1)^{m(s+1)}.$$

Here $G_j \simeq GL(n(j))$ stands for the obvious j-th block of $L(\tilde{\pi})$. Set

$$C_\lambda^{(j)} := \begin{cases} \mathrm{Ind}(\tilde{P}(\pi) \cap G_1 \uparrow G_1)([\lambda^{(0)}] \otimes [\Lambda_1]^{m(1)} \otimes [\nu^{(1)}]) & (j=1) \\ \mathrm{Ind}(\tilde{P}(\pi) \cap G_j \uparrow G_j)([\Lambda_j]^{m(j)} \otimes [\nu^{(j)}]) & (2 \le j \le s). \end{cases}$$

Then appealing to induction by stages, we have

$$C_\lambda = \mathrm{Ind}(\tilde{P}(\tilde{\pi}) \uparrow Sp(n)) \left((\bigotimes_{j=1}^{s} C_\lambda^{(j)}) \otimes [\nu^{(s+1)}] \otimes [\Lambda_{s+1}]^{m(s+1)} \right).$$

As $C_\lambda^{(j)}$ is an irreducible bimodule for G_j due to Barbasch and Vogan ([30] Proposition 12.2), each Harish-Chandra bimodule map for G_j

$$U(\mathfrak{g}_1)/I(\lambda^{(0)}, \Lambda_1, \ldots, \Lambda_1, \nu^{(1)}) \to C_\lambda^{(1)}$$
$$U(\mathfrak{g}_j)/I(\Lambda_j, \ldots, \Lambda_j, \nu^{(j)}) \to C_\lambda^{(j)} \quad (2 \le j \le s)$$

is surjective. From Theorem 5.5 of [33] combined with the exactness of the functor $\mathrm{Ind}(\tilde{P}(\tilde{\pi}) \uparrow Sp(n))$, the bimodule map

$$U(\mathfrak{sp}(n))/I(\gamma) \to C_\lambda$$

is also surjective (γ is defined in (6.3.2)(a)). Therefore C_λ is generated by a unique (up to scalar) $Sp(n)$-fixed vector as a Harish-Chandra bimodule for $Sp(n)$.

The similar argument is available for $\lambda + \xi$. More precisely, one should replace $\tilde{\pi}$ by its refinement according to $n(1) = N + (m(1) + \Lambda_0 - \Lambda_1)$. Thus we have shown the condition (6.5.4)(b). □

6.10. verification of (6.5.4)(c)

Retain notations in Theorem(6.3.1) and §6.6-7. Suppose we are in the setting of (6.7.1-4). Define another ordered partition of $n = N + r + l$ by

$$\begin{aligned}\pi' &= (N, 1^{m(1)}, \dots, 1^{m(s+1)}, \Lambda_0 - \Lambda_1, \dots, \Lambda_s - \Lambda_{s+1}) \\ &= (N, 1^r, \Lambda_0 - \Lambda_1, \dots, \Lambda_s - \Lambda_{s+1}),\end{aligned}$$

a permutation of π (Definition(6.7.5)). We define a Harish-Chandra bimodule for $L(\pi')$ by

$$(\sigma')_\lambda := [\lambda^{(0)}] \otimes (\bigotimes_{j=1}^{s+1} [\Lambda_j]^{m(j)}) \otimes (\bigotimes_{j=1}^{s+1} [\nu^{(j)}]),$$

and a Harish-Chandra bimodule for $Sp(n)$ by

$$(C')_\lambda := \operatorname{Ind}(\tilde{P}(\pi') \uparrow Sp(n))((\sigma')_\lambda).$$

We will construct Harish-Chandra bimodule maps for $Sp(n)$:

$$C_\lambda \to (C')_\lambda \tag{6.10.1}$$

$$(C')_\lambda \to R_\gamma(\mathfrak{l} : \mathfrak{g}) \tag{6.10.2}$$

that are non-zero on $Sp(n)$-fixed vectors.

As for the first map (6.10.1), it is proved on $GL(n)$ level. In fact, put

$$\tilde{\gamma} := ({}^t\lambda^{(0)}, \Lambda_1^{m(1)}, {}^t\nu^{(1)}, \dots, \Lambda_s^{m(s)}, {}^t\nu^{(s)}, {}^t\nu^{(s+1)}, \Lambda_{s+1}^{m(s+1)}) \in \mathbb{C}^n,$$

$$\tilde{\gamma}' := ({}^t\lambda^{(0)}, \Lambda_1^{m(1)}, \dots, \Lambda_s^{m(s)}, \Lambda_{s+1}^{m(s+1)}, {}^t\nu^{(1)}, \dots, {}^t\nu^{(s)}, {}^t\nu^{(s+1)}) \in \mathbb{C}^n,$$

Then there is a commutative diagram of intertwining operators between Harish-Chandra bimodules (see Definition(6.6.2)):

$$\begin{array}{ccc} \mathrm{Ind}(B \uparrow GL(n))(R_{\tilde{\gamma}}(\mathfrak{h}:\mathfrak{h})) & \longrightarrow & \mathrm{Ind}(B \uparrow GL(n))(R_{\tilde{\gamma}'}(\mathfrak{h}:\mathfrak{h})) \\ \cup & & \cup \\ \mathrm{Ind}(P(\pi) \uparrow GL(n))(\sigma_\lambda) & \longrightarrow & \mathrm{Ind}(P(\pi') \uparrow GL(n))((\sigma')_\lambda). \end{array}$$

The point here is that these horizontal maps are non-zero on $GL(n)$-fixed vectors ([28] Chapter 4, see also [30] Theorem 11.5). By inducing it to $Sp(n)$, the induced Harish-Chandra bimodule map

$$\begin{aligned} C_\lambda &\simeq \mathrm{Ind}(\tilde{P}((n)) \uparrow Sp(n))\,\mathrm{Ind}(P(\pi) \uparrow GL(n))(\sigma_\lambda) \\ &\to (C')_\lambda \simeq \mathrm{Ind}(\tilde{P}((n)) \uparrow Sp(n))\,\mathrm{Ind}(P(\pi') \uparrow GL(n))((\sigma')_\lambda) \end{aligned}$$

is non-zero on the $Sp(n)$-fixed vector.

As for the second map (6.10.2) we use induction by stages according to

$$L(\pi') \subset L_{\mathbb{C}} := GL(N) \times GL(1)^r \times Sp(l) \subset Sp(n).$$

Then the desired result is obtained from the corresponding one for $L_{\mathbb{C}}$, that is,

$$\mathrm{Ind}(\tilde{P}(\pi') \cap L_{\mathbb{C}} \uparrow L_{\mathbb{C}})((\sigma')_\lambda) \to R_\gamma(\mathfrak{l}:\mathfrak{l})$$

is a surjective Harish-Chandra bimodule map for $L_{\mathbb{C}}$.

The argument for $\lambda + \xi$ is quite similar.

6.11. proof of Corollary(6.4.1)

Suppose we are in the setting of Corollary(6.4.1). Define the parabolic subalgebra $\mathfrak{q} = \mathfrak{l} + \mathfrak{u}$ of $\mathfrak{g}$ such that $\mathfrak{q} \subset \mathfrak{p}$, $\mathfrak{h} \subset \mathfrak{l} \subset \mathfrak{m}$ and $\mathfrak{l} \simeq \mathfrak{gl}(n_1) \oplus \mathfrak{gl}(1)^{n-n_1-l} \oplus \mathfrak{sp}(l)$.

We proceed by induction on n. Put $\mathfrak{g}' := \mathfrak{gl}(n_1) \oplus \mathfrak{sp}(n-n_1)$, $s' := \dim(\mathfrak{n} \cap \mathfrak{k} \cap \mathfrak{g}')$. We will show that $\mathcal{R}_{\mathfrak{p}}^s(\mathbb{C}_\lambda)$ is irreducible as $U(\mathfrak{g})$-module assuming the irreducibility in the $\mathfrak{sp}(n-n_1)$ case. If $\gamma_{n_1} \geq l$, then γ is weakly good with respect to $\mathfrak{p}+\mathfrak{g}' \subset \mathfrak{g}$ and so the inductive assumption assures that $\left(\mathcal{R}_{\mathfrak{p}}^{\mathfrak{g}}\right)^s(\mathbb{C}_\lambda) \simeq \left(\mathcal{R}_{\mathfrak{p}+\mathfrak{g}'}^{\mathfrak{g}}\right)^{s-s'}\left(\left(\mathcal{R}_{\mathfrak{p}\cap\mathfrak{g}'}^{\mathfrak{g}'}\right)^{s'}(\mathbb{C}_\lambda)\right)$ is irreducible (or zero) (Fact(1.4.1)(1), Lemma(3.2.1)). Thus we shall concentrate on the case where $\gamma_{n_1} < l$.

First recall that our assumption on $\gamma = \lambda + \rho_{\mathfrak{m}}$ implies that γ is in the weakly fair range with respect to both $\mathfrak{q} \subset \mathfrak{g}$ and $\mathfrak{p} \subset \mathfrak{g}$. We have a surjective Harish-Chandra bimodule map for $M_{\mathbb{C}}$ (cf. (6.6.1)):

$$\operatorname{Ind}(M_{\mathbb{C}} \cap Q_{\mathbb{C}} \uparrow M_{\mathbb{C}})\,(R_\gamma(\mathfrak{l}:\mathfrak{l})) \to R_\gamma(\mathfrak{m}:\mathfrak{m}).$$

Applying the exact functor $\operatorname{Ind}(P_{\mathbb{C}} \uparrow G_{\mathbb{C}})$ with trivial $\mathfrak{n}$ action, we have a surjective Harish-Chandra bimodule map for $G_{\mathbb{C}}$:

$$R_\gamma(\mathfrak{l}:\mathfrak{g}) \to R_\gamma(\mathfrak{m}:\mathfrak{g}).$$

Here we used $R_\gamma(\mathfrak{l}:\mathfrak{g}) \simeq \operatorname{Ind}(P_{\mathbb{C}} \uparrow G_{\mathbb{C}})\operatorname{Ind}(M_{\mathbb{C}} \cap Q_{\mathbb{C}} \uparrow M_{\mathbb{C}})\,(R_\gamma(\mathfrak{l}:\mathfrak{l}))$. Thus the $U(\mathfrak{g})$-action on $\mathcal{R}_{\mathfrak{p}}^s(\mathbb{C}_\lambda)$ factors through algebra homomorphisms

$$U(\mathfrak{g}) \to R_\gamma(\mathfrak{l}:\mathfrak{g}) \twoheadrightarrow R_\gamma(\mathfrak{m}:\mathfrak{g}).$$

Now we apply Theorem(6.3.1) to $r = n_2 + \cdots + n_k$, $N = n_1$ and $M = \mathcal{R}_{\mathfrak{p}}^s(\mathbb{C}_{\lambda+\xi})$. (As γ is a $\mathfrak{Z}(\mathfrak{m})$-infinitesimal character of a one dimensional representation, γ is of the form (6.3.2)(a) with (6.3.2)(b).) Choose ξ as in Theorem(6.3.1), then $\mathcal{R}_{\mathfrak{p}}^s(\mathbb{C}_{\lambda+\xi})$ is irreducible by the inductive assumption as we saw, in the case $\gamma_{n_1} \geq l$. Therefore if we show

$$\psi_{\gamma+\xi}^{\gamma}\mathcal{R}_{\mathfrak{p}}^s(\mathbb{C}_{\lambda+\xi}) = \mathcal{R}_{\mathfrak{p}}^s(\mathbb{C}_\lambda), \tag{6.11.1}$$

then we are done. From Lemma(3.4.1)(2), it suffices to show

$$\mu = -\xi,$$

if $\mu \in \mathfrak{h}^* \simeq \mathbb{C}^n$ is a weight in $F(Sp(n), -\xi)$ satisfying

$$\begin{cases} \mu \text{ is dominant for } \Delta^+(\mathfrak{m}, \mathfrak{h}), \\ \gamma + \xi + \mu = w \cdot \gamma \quad \text{for some } w \in W(\mathfrak{sp}(n)). \end{cases}$$

This is derived from Lemma(6.8.1) by the following observation: With notation in Lemma(6.8.1), reorder π, γ as

$$\pi' := (N, 1^{m(1)+\cdots+m(s)+m(s+1)}, \Lambda_0 - \Lambda_1, \ldots, \Lambda_{s-1} - \Lambda_s, \Lambda_s - 1),$$
$$\gamma' := (\lambda^{(0)}, \lambda_1, \ldots, \lambda_r, \nu^{(1)}, \ldots, \nu^{(s)}, \nu^{(s+1)}).$$

The same statement for γ', $\mathfrak{l}(\pi')$ as in Lemma(6.8.1) holds (use an inner automorphism of $Sp(n)$) and its assumption is satisfied under (6.11.2) because $\mathfrak{l}(\pi') \subset \mathfrak{m}$. (Recall that $(\nu^{(1)}, \nu^{(2)}, \ldots, \nu^{(s)}, \nu^{(s+1)})$ is a grouping of $(l, l-1, \ldots, 2, 1)$.) Hence (6.11.1) holds and Corollary(6.4.1) is proved.

§7. Proof of vanishing results outside the fair range

This section is devoted to Part(2) of Theorems 1 and 2 (see §2): the vanishing result, when $j \neq S \equiv \dim(\mathfrak{u} \cap \mathfrak{k})$, for the derived functor modules $\mathcal{R}^j_{\mathfrak{q}}(\mathbb{C}_\lambda)$ with small parameter λ such that $\mathbb{C}_\lambda$ lies outside the fair range. When $G = SO_0(p,q)$, the condition (2.6.2) implies that $\mathbb{C}_\lambda$ (resp. $\mathbb{C}_{\lambda'}$) lies in the weakly fair range with respect to $\mathfrak{q}$ (resp. $\mathfrak{q}'$). Then Part(2) of Theorem 3 is a direct consequence of Fact(1.4.2)(1-a). When $G = U(p,q)$, we can apply Lemma(3.4.1)(2) and Fact(1.4.2)(1-a). This is fairly easy and we shall give a preliminary Lemma(7.1.1). The rest of this section (§7.2-4) will be devoted to the case $G = Sp(p,q)$. The non-trivial part there is only the crossing of the wall $\lambda_r = 0$.

7. proof in complex case

When $G = U(p,q)$, the iteration of the following lemma and Lemma(3.4.1)(2) reduces the vanishing of $\mathcal{R}^j_{\mathfrak{q}}(\mathbb{C}_\lambda)$ $(j \neq S)$ under the condition (2.4.2) to that under the condition

$$\lambda_1 > \lambda_2 > \cdots > \lambda_r > Q \geq \lambda_{r+1} \geq \cdots \geq \lambda_{r+s}.$$

Similarly, this is then reduced to the vanishing under the condition

$$\lambda_1 > \lambda_2 > \cdots > \lambda_r > Q > -Q > \lambda_{r+1} > \cdots > \lambda_{r+s}.$$

Now the vanishing of $\mathcal{R}^j_{\mathfrak{q}}(\mathbb{C}_\lambda)$ $(j \neq S)$ in this last case is well known (see Fact(1.4.1)(1-a) or Fact(1.4.2)(1-a)). So it suffices to show

Lemma(7.1). *Let $G = U(p,q)$ and $\mathfrak{q} = \mathfrak{q}(r,s) = \mathfrak{l} + \mathfrak{u}$ $(r + s \leq p)$ be a θ-stable parabolic subalgebra. Retain notations as in §2.3. Let $\lambda = (\lambda_1, \ldots, \lambda_{r+s}) \in \mathfrak{t}^*$. Fix*

$i\,(1 \le i \le r+s)$ and set $\lambda' := \lambda + f_i \in \mathfrak{t}^*$. Assume

(7.1.2)(a) $\qquad \lambda_i \ge -Q.$

(7.1.2)(b) $\qquad \lambda_j \ne \lambda_i + 1$ *for any* j $(1 \le j \le r+s)$.

Then $\mathcal{A}(\lambda' \triangleright \lambda) = \{\lambda\}$ (see §3.3 for definition).

Proof. Fix a positive system $\Delta^+(\mathfrak{l},\mathfrak{h})$ so that $\rho_{\mathfrak{l}} = (\overbrace{0,\ldots,0}^{r+s}, Q, Q-1, \ldots, -Q)$. We shall prove $\mathcal{A}(\Delta^+(\mathfrak{l},\mathfrak{h}); \lambda' \triangleright \lambda) = \{\lambda\}$. (Then we tell *a fortiori* $\mathcal{A}(\Delta^+(\mathfrak{l},\mathfrak{h}); \lambda' \triangleright \lambda)$ is independent of a particular choice of $\Delta^+(\mathfrak{l},\mathfrak{h})$. This is the reason why we omit $\Delta^+(\mathfrak{l},\mathfrak{h})$ in the statement of Lemma(7.1.1).)

Suppose $\mu = (\mu_1, \ldots, \mu_{p+q}) \in \mathcal{A}(\Delta^+(\mathfrak{l},\mathfrak{h}); \lambda' \triangleright \lambda)$. As $\mu - \lambda'$ is a weight of a $p+q$ dimensional representation $F(\mathfrak{gl}(p+q,\mathbb{C}), -f_i) = F(\mathfrak{gl}(p+q,\mathbb{C}), -f_1)$, we can find $k\,(1 \le k \le p+q)$ such that

(7.1.3) $\qquad \mu = \lambda' - f_k = \lambda + f_i - f_k.$

Thus $\mu - \lambda'$ is dominant for $\Delta^+(\mathfrak{l},\mathfrak{h})$ iff $1 \le k \le r+s$ or $k = p+q$.

Let us observe that $k \ne p+q$. If μ were to be $\lambda + f_i - f_{p+q}$, then (7.1.2)(a) and (7.1.3) would imply

$$\#\{j\,;\,(\mu+\rho_{\mathfrak{l}})_j = -Q-1\} = \#\{j\,;\,(\lambda+\rho_{\mathfrak{l}})_j = -Q-1\} + 1,$$

contradicting to the fact that $\mu + \rho_{\mathfrak{l}} \in \mathfrak{S}_{p+q} \cdot (\lambda + \rho_{\mathfrak{l}})$. Therefore we have $1 \le k \le r+s$ and $\mu = \lambda + f_i - f_k \in \mathfrak{t}^*$. Then $\mu + \rho_{\mathfrak{l}} \in \mathfrak{S}_{p+q} \cdot (\lambda + \rho_{\mathfrak{l}})$ implies $\mu \in \mathfrak{S}_{r+s} \cdot \lambda$. Now (7.1.2)(b) assures $k = i$ and $\mu = \lambda + f_i - f_i = \lambda$. Hence $\mathcal{A}(\Delta^+(\mathfrak{l},\mathfrak{h}); \lambda' \triangleright \lambda) = \{\lambda\}$. □

7.2. vanishing result in quaternionic case

The rest of this section will be devoted to proving Theorem 1 (2). That is,

Theorem 7.2.1. *Let $G = Sp(p,q)$ and fix an integer r $(1 \le r \le p)$. Set $Q := p+q-r$ and let $\mathfrak{q} \equiv \mathfrak{q}(r)$ be a θ-stable parabolic subalgebra as in §2.1. Retain the notation there. If $\lambda = (\lambda_1, \dots, \lambda_r) \in \mathbb{Z}^r \subset \mathfrak{t}^*$ satisfies*

$$\lambda_1 \ge \lambda_2 \ge \cdots \ge \lambda_{r-1} \ge |\lambda_r|, \ \lambda_r \ge -Q, \tag{7.2.2}$$

then $\mathcal{R}_{\mathfrak{q}}^j(\mathbb{C}_\lambda) = 0$ for any $j \ne S$.

Remark 7.2.3. The assumption $\lambda_r \ge -Q$ in (7.2.2) is crucial. In fact, it does happen that $\mathcal{R}_{\mathfrak{q}}^{S-i}(\mathbb{C}_\lambda) \ne 0$ $(i = 0,1)$ when $\lambda_r < -Q$. (I checked this when $p = r$ by using Vogan's U_α calculus.)

7.3. maximal parabolic case

First we prove Theorem 7.2.1 in the case where $r = 1$ so that $\mathfrak{q}$ is a maximal parabolic subalgebra in $\mathfrak{g}$. This is a main part in the proof.

Lemma 7.3.1. *Theorem*(7.2.1) *holds when $r = 1$.*

Proof. We write $\lambda(n) := n \in \mathfrak{t}^* \simeq \mathbb{C}$ for $n \in \mathbb{Z}$. Then $\mathbb{C}_{\lambda(n)}$ is in the weakly fair range iff $n \ge 0$. Because of the vanishing theorem, $\mathcal{R}_{\mathfrak{q}}^j(\mathbb{C}_{\lambda(n)}) = 0$ $(j \ne S)$, in the weakly fair range (Fact(1.4.2)(1-a)), we should concentrate our attention on $0 > n \ge -Q$. Then thanks to Lemma(3.4.1), it suffices to treat only the case where $n = -1$ because

$$\mathcal{A}(\lambda(-i) \triangleright \lambda(-i-1)) = \{\lambda(-i-1)\} \qquad (-1 \ge i \ge -Q+1).$$

(See §3.3 for notation.) Let us prove the vanishing result for $\lambda = \lambda(-1)$. For this, we apply Lemma(3.4.1)(3-a,b) and Claim(3.4.7). But we give a little detailed explanation of it for the benefit of the reader. First observe that

$$\mathcal{A}(\lambda(1) \triangleright \lambda(0)) = \mathcal{A}(\lambda(-1) \triangleright \lambda(0)) = \{\lambda(0)\}, \tag{7.3.2)(a}$$

$$\mathcal{A}(\lambda(0) \triangleright \lambda(1)) = \mathcal{A}(\lambda(0) \triangleright \lambda(-1)) = \{\lambda(1), \lambda(-1)\}, \tag{7.3.2)(b}$$

and that $\mathcal{R}^j_{\mathfrak{q}}(\mathbb{C}_{\lambda(1)}) = \mathcal{R}^j_{\mathfrak{q}}(\mathbb{C}_{\lambda(0)}) = 0$ for $j \neq S$. From (7.3.2)(a), we have

$$\psi^{\lambda(0)+\rho_{\mathfrak{l}}}_{\lambda(1)+\rho_{\mathfrak{l}}}(\mathcal{R}^j_{\mathfrak{q}}(\mathbb{C}_{\lambda(1)})) = \mathcal{R}^j_{\mathfrak{q}}(\mathbb{C}_{\lambda(0)}),$$
$$\psi^{\lambda(0)+\rho_{\mathfrak{l}}}_{\lambda(-1)+\rho_{\mathfrak{l}}}(\mathcal{R}^j_{\mathfrak{q}}(\mathbb{C}_{\lambda(-1)})) = \mathcal{R}^j_{\mathfrak{q}}(\mathbb{C}_{\lambda(0)}).$$

Applying Claim(3.4.7) with $F = F(Sp(p,q),\lambda(1)) \, (\simeq \mathbb{C}^{2p+2q})$, we have a filtration of $\mathfrak{q}$-modules:

$$\{0\} = F_0 \subset F_1 \subset F_2 \subset F_3 = F$$

such that $F_1 \simeq \mathbb{C}_{\lambda(1)}$, $F_2/F_1 \simeq 1 \boxtimes F(Sp(p-1,q),(1,0,\dots,0))$ and $F/F_2 \simeq \mathbb{C}_{\lambda(-1)}$. From (7.3.2)(b) and Lemma(3.4.1)(3-b)$'$, we have a long exact sequence of $(\mathfrak{g},K)$-modules:

$$0 \to \mathcal{R}^{S-1}_{\mathfrak{q}}(\mathbb{C}_{\lambda(-1)}) \to \mathcal{R}^S_{\mathfrak{q}}(\mathbb{C}_{\lambda(1)}) \to \psi^{\lambda(1)+\rho_{\mathfrak{l}}}_{\lambda(0)+\rho_{\mathfrak{l}}}(\mathcal{R}^S_{\mathfrak{q}}(\mathbb{C}_{\lambda(0)})) \to \mathcal{R}^S_{\mathfrak{q}}(\mathbb{C}_{\lambda(-1)}) \to 0,$$

and $\mathcal{R}^j_{\mathfrak{q}}(\mathbb{C}_{\lambda(-1)}) = 0$ for $j \neq S-1, S$. Since $\mathcal{R}^S_{\mathfrak{q}}(\mathbb{C}_{\lambda(1)})$ is (nonzero) irreducible by Corollary(6.4.1), we have either

$$\mathcal{R}^{S-1}_{\mathfrak{q}}(\mathbb{C}_{\lambda(-1)}) = 0,$$

or

$$\mathcal{R}^{S-1}_{\mathfrak{q}}(\mathbb{C}_{\lambda(-1)}) \simeq \mathcal{R}^S_{\mathfrak{q}}(\mathbb{C}_{\lambda(1)}).$$

But the latter case is impossible. Indeed, applying $\psi^{\lambda(0)+\rho_{\mathfrak{l}}}_{\lambda(1)+\rho_{\mathfrak{l}}} = \psi^{\lambda(0)+\rho_{\mathfrak{l}}}_{\lambda(-1)+\rho_{\mathfrak{l}}}$ to the latter (false) isomorphism, we would have $\mathcal{R}^{S-1}_{\mathfrak{q}}(\mathbb{C}_{\lambda(0)}) = \mathcal{R}^S_{\mathfrak{q}}(\mathbb{C}_{\lambda(0)})$. Since $\mathbb{C}_{\lambda(0)}$ is in the weakly fair range, we have $\mathcal{R}^{S-1}_{\mathfrak{q}}(\mathbb{C}_{\lambda(0)}) = 0$, while $\mathcal{R}^S_{\mathfrak{q}}(\mathbb{C}_{\lambda(0)}) \neq 0$ by Corollary(4.3.6). This is a contradiction. Therefore $\mathcal{R}^j_{\mathfrak{q}}(\mathbb{C}_{\lambda(-1)}) = 0$ for all $j \neq S$. This completes the proof. □

7.4. general parabolic case

Proof of Theorem(7.2.1). Fix $\lambda = (\lambda_1, \dots, \lambda_r)$ satisfying (7.2.2). The spectral sequence of induction by stages (Lemma(3.2.1)) corresponding to

$$L := \mathbb{T}^r \times Sp(p-r, q) \subset M := \mathbb{T}^{r-1} \times Sp(p-r+1, q) \subset G = Sp(p,q)$$

collapses to

$$\left(\mathcal{R}^{\mathfrak{g}}_{\mathfrak{q}(r-1)}\right)^{i} \left(\mathcal{R}^{\mathfrak{m}}_{\mathfrak{q}\cap\mathfrak{m}}\right)^{S'} (\mathbb{C}_\lambda) \simeq \mathcal{R}^{i+S'}_{\mathfrak{q}}(\mathbb{C}_\lambda), \tag{7.4.1}$$

from the vanishing result Lemma(7.3.1) applied to a maximal parabolic subalgebra $\mathfrak{q} \cap \mathfrak{m} \subset \mathfrak{m}$ and $Q = (p-r+1) + q - 1 = p + q - r$. Here $S' = \dim(\mathfrak{u} \cap \mathfrak{k} \cap \mathfrak{m})$. Take a weight $\mu = (\mu_1, \dots, \mu_{r-1}, \lambda_r) \in \mathbb{Z}^r \subset \mathfrak{t}^*$ such that

$$\mu_1 > \mu_2 > \cdots > \mu_{r-1} > \max(|\lambda_r|, Q).$$

Since $\mathcal{R}^{S'}_{\mathfrak{q}\cap\mathfrak{m}}(\mathbb{C}_\mu)$ is in the good range with respect to $\mathfrak{q}(r-1) \subset \mathfrak{g}$, we have $\mathcal{R}^j_{\mathfrak{q}}(\mathbb{C}_\mu) = 0$ for $j \neq S$. We can easily find a sequence $\lambda^{(0)} = \mu, \lambda^{(1)}, \dots, \lambda^{(n)} = \lambda$ so that

$$\mathcal{A}\left(\lambda^{(i-1)} \triangleright \lambda^{(i)}\right) = \{\lambda^{(i)}\} \text{ for } i = 1, 2, \dots, n.$$

Hence an inductive argument shows $\mathcal{R}^j_{\mathfrak{q}}(\mathbb{C}_\lambda) = 0$ for all $j \neq S$ from Lemma(3.4.1). □

§8. Proof of the inequivalence results

In this section we shall show the results on pairwise inequivalence among $\mathcal{R}_{\mathfrak{q}}^{S}(\mathbb{C}_\lambda)$'s, i.e. Part (4) of the Theorems in §2. Our method is based on the K-spectrum calculated in §4. In view of the $\mathfrak{Z}(\mathfrak{g})$-infinitesimal characters, the non-trivial cases are when $G = Sp(p,q)$ or when $G = SO_0(p,q)$ with $p = 2r$.

8. quarternionic case

Suppose we are in the setting of §2.2. It suffices to compare the modules with the same $\mathfrak{Z}(\mathfrak{g})$-infinitesimal character to prove Part (4) of Theorem 1. That is, we must show:

Proposition 8.1. *Let $G = Sp(p,q)$ and fix an integer r $(1 \le r \le p)$. Retain the notation in §2.1. Let $\lambda = (\lambda_1, \dots, \lambda_{r-1}, \lambda_r) \in \mathbb{Z}^r \subset \mathfrak{t}^*$ satisfy*

$$(8.2)\qquad \begin{cases} \lambda_1 > \lambda_2 > \cdots > \lambda_{r-1} > \lambda_r > 0, \\ \lambda_{r-2q} \ge Q + 1 (\equiv p + q - r + 1) \quad \text{when } r > 2q. \end{cases}$$

If we set $\lambda' = (\lambda_1, \dots, \lambda_{r-1}, -\lambda_r)$, then $\mathcal{R}_{\mathfrak{q}}^{S}(\mathbb{C}_\lambda) \not\simeq \mathcal{R}_{\mathfrak{q}}^{S}(\mathbb{C}_{\lambda'})$ as a $(\mathfrak{g}, K)$-module.

Proof. Let l $(0 \le l \le r-1)$ be an integer such that $\lambda_l > Q \ge \lambda_{l+1}$. We may assume $\lambda_l >> Q$ so that $\lambda_l - p + q + l - 1 \ge 0$. We may also assume $Q \ge \lambda_r$. This is possible because the general statement is derived from the corresponding one in this special case by reduction to absurdity (use the translation principle). In our present situation, $\mathcal{R}_{\mathfrak{q}}^{i}(\mathbb{C}_\lambda) = \mathcal{R}_{\mathfrak{q}}^{i}(\mathbb{C}_{\lambda'}) = 0$ when $i \ne S$ by Theorem(7.2.1). Hence, it suffices to find $\mu \in \mathfrak{t}^*$ such that $M(\mathfrak{q}, \lambda, \mu) \ne M(\mathfrak{q}, \lambda', \mu)$ (see §4.1 for notation). We define

$b, b' \in \mathbb{Z}^r$ by (4.3.1) and $k, k' \in \mathbb{N}$ by (4.3.4) corresponding to λ, λ' respectively. Set $\delta = (b_1, \dots, b_k, 0, \dots, 0)$, $\delta' = (b'_1, \dots, b'_{k'}, 0, \dots, 0) \in \mathbb{Z}^r$. Let us show

$$M(\mathfrak{q}, \lambda, \delta') < M(\mathfrak{q}, \lambda', \delta').$$

As $b_l = \lambda_l - p + q + l - 1 \geq 0$ from our assumption, we have $k, k' \geq l$ and $Q \geq \lambda_{k+1}, \lambda'_{k'+1}$ respectively. We divide now into two cases according as $\delta = \delta'$ or not.

(I) $\delta \neq \delta'$.

In this case we have $\sum_{i=1}^r \delta'_i < \sum_{i=1}^r \delta_i = \sum_{i=1}^k b_i$. Hence, $M(\mathfrak{q}, \lambda, \delta') = 0 < M(\mathfrak{q}, \lambda', \delta')$ from Proposition(4.3.2)(2)-(3) (see also Remark(4.3.5)).

(II) $\delta = \delta'$.

In this case we have $k = k'$. Combining Proposition(4.3.2)(2) with the explicit formula of $d(n, l; x)$ in Lemma(4.2.6) and putting $P := p - q + 1$, then we have

$$\begin{aligned} \frac{M(\mathfrak{q}, \lambda, \delta)}{M(\mathfrak{q}, \lambda', \delta')} &= \frac{d(2q, r-k; -\lambda_{k+1} + P, \dots, -\lambda_{r-1} + P, -\lambda_r + P)}{d(2q, r-k; -\lambda_{k+1} + P, \dots, -\lambda_{r-1} + P, \lambda_r + P)} \\ &= \frac{\prod_{j=1}^{2q-r+k} (j - \lambda_r + p - q) \prod_{k+1 \leq j < r} (\lambda_j - \lambda_r)}{\prod_{j=1}^{2q-r+k} (j + \lambda_r + p - q) \prod_{k+1 \leq j < r} (\lambda_j + \lambda_r)} \\ &< 1. \end{aligned}$$

Hence $M(\mathfrak{q}, \lambda, \delta') = M(\mathfrak{q}, \lambda, \delta) < M(\mathfrak{q}, \lambda', \delta')$. □

8.2. orthogonal case

Suppose we are in the setting of §2.6 and §4.5. Assume $p = 2r$ and (2.6.3). Put

$$\begin{aligned} \delta &:= \mu_{\lambda|\mathfrak{t}^c} = (\lambda_1 + \frac{q}{2} - r + 1, \dots, \lambda_{r-1} + \frac{q}{2} - 1, \lambda_r + \frac{q}{2}) \in (\mathfrak{t}_1^c)^*, \\ \delta' &:= \mu'_{\lambda'|\mathfrak{t}^c} = (\lambda_1 + \frac{q}{2} - r + 1, \dots, \lambda_{r-1} + \frac{q}{2} - 1, -\lambda_r - \frac{q}{2}) \in (\mathfrak{t}_1^c)^*, \end{aligned}$$

Then $\delta \neq \delta'$ and

$$M(\mathfrak{q},\lambda,\delta) = M(\mathfrak{q}',\lambda',\delta') = 1$$
$$M(\mathfrak{q},\lambda,\delta') = M(\mathfrak{q}',\lambda',\delta) = 0$$

This follows from Proposition(4.5.2), but is easy to check directly because μ_λ, $\mu'_{\lambda'}$ are $\Delta^+(\mathfrak{k})$ dominant in this case. Since $\mathcal{R}_{\mathfrak{q}}^{S-j}(\mathbb{C}_\lambda) = \mathcal{R}_{\mathfrak{q}'}^{S-j}(\mathbb{C}_{\lambda'}) = 0$ when $j \neq 0$, we have $\mathcal{R}_{\mathfrak{q}}^{S}(\mathbb{C}_\lambda) \not\simeq \mathcal{R}_{\mathfrak{q}'}^{S}(\mathbb{C}_{\lambda'})$.

References

[1] J.Adams, *Unitary highest weight modules*, Adv. in Math. **63** (1987), 113-137.

[2] F.Bien, *Spherical $\mathcal{D}$-modules and representations of reductive Lie groups*, Ph.D. dissertation, M.I.T., Cambridge, Massachusetts (1986).

[3] W.Borho and J-L.Brylinski, *Differential operators on homogeneous spaces* I, Invent.Math. **69** (1982), 437-476.

[4] J.T.Chang, *Remarks on localization and standard modules: The duality theorem on a generalized flag variety*, preprint, M.S.R.I., Berkeley.

[5] D.Collingwood, *Representations of rank one Lie groups*, Research notes in Math., Pitman Publishing, London, 1985.

[6] M.Duflo, *Sur la classification des ideaux primitifs dans l'algebre enveloppante d'une algebre de Lie semi-simple*, Ann. of Math. **105** (1977), 107-120.

[7] T.J.Enright, R.Parthasarathy, N.R.Wallach, and J.A.Wolf, *Unitary derived functor module with small spectrum*, Acta.Math. **154** (1985), 105-136.

[8] M.Flensted-Jensen, *Discrete series for semisimple symmetric spaces*, Ann. of Math. **111** (1980), 253-311.

[9] ———, *Analysis on non-Riemannian symmetric spaces*, Conference Board, No.61, A.M.S., 1986.

[10] S.Helgason, *A duality for symmetric spaces with applications to group representations*, Adv. in Math. **5** (1970), 1-154.

[11] H.Hecht, D.Miličic, W.Schmid and J.A.Wolf, *Localization and standard modules for real semisimple Lie groups*, Invent.Math. **90** (1987), 297-332.

[12] W.Hesselink, *Polarizations in the classical groups*, Math.Z. **160** (1978), 217-234.

[13] T.Kobayashi, *Unitary representations realized in sections of vector bundles over semisimple symmetric spaces* (in Japanese), Master's dissertation I, Univ.of Tokyo (1987).

[14] ———, *Construction of discrete series for vector bundles over semisimple symmetric spaces* (in Japanese), RIMS Kyoto Kokyuroku **642** (1988), 134-156.

[15] B.Kostant, *Lie algebra cohomology and the generalized Borel-Weil theorem*, Ann. of Math. **74** (1961), 329-387.

[16] ———, *Irreducibility of certain series of representations*, Publication of 1971 Summer School in Math., (ed. I.M.Gel'fand), Bolyai-Janos Math.Soc., Budapest.

[17] H.Kraft and C.Procesi, *Closures of conjugacy classes of matrices are normal*, Invent.Math. **53** (1979), 227-247.

[18] ———, *On the geometry of conjugacy classes in the classical groups*, Comment.Math.Helv. **57** (1982), 539-602.

[19] A.Magnus, *Non-spherical principal series representations of a semisimple Lie group*, Memoirs No.216, A.M.S., 1979.

[20] T.Matsuki and T.Oshima, *A description of discrete series for semisimple symmetric spaces*, Advanced Studies in Pure Math. **4** (1984), 331-390.

[21] T.Matsuki, *A description of discrete series for semisimple symmetric spaces* II, Advanced Studies in Pure Math. **14** (1988), 531-540.

[22] T.Oshima, *Asymptotic behavior of spherical functions on semisimple symmetric spaces*, Advanced Studies in Pure Math. **14** (1988), 561-601.

[23] H.Schlichtkrull, *A series of unitary irreducible representations induced from a symmetric subgroup of a semisimple Lie group*, Invent.Math. **68** (1982), 497-516.

[24] ———, *The Langlands parameters of Flensted-Jensen's discrete series for semisimple symmetric spaces*, J.Funct.Anal. **50** (1983), 133-150.

[25] ———, *One dimensional K-types in finite dimensional representations of semisimple Lie groups. A generalization of Helgason's theorem*, Math.Scand. **54** (1984), 279-294.

[26] N.Shimeno, *Eigenspaces of invariant differential operators on a homogeneous line bundle on a Riemannian symmetric space*, to appear in J.Fac.Sci.Univ.Tokyo, **37-1**.

[27] D.Vogan, *Irreducible characters of semisimple Lie groups.* I, Duke Math.J. **46** (1979), 61-108.

[28] ———, *Representations of real reductive Lie groups*, Birkhäuser, 1981.

[29] ———, *Unitarizability of certain series of representations*, Ann. of Math. (1984), 141-187.

[30] ———, *The unitary dual of $GL(n)$ over an archimedean field*, Invent.Math. (1986), 449-505.

[31] ———, *The orbit method and primitive ideals for semisimple Lie algebras*, Lie algebras and related topics, vol. 5, CMS Conference Proceedings, 1986, pp. 281-316.

[32] ———, *Unitary representations of reductive Lie groups*, Princeton University Press, Princeton, New Jersey, 1987.

[33] ———, *Irreducibility of discrete series representations for semisimple symmetric spaces*, Advanced Studies in Pure Math. **14** (1988), 191-221.

[34] N.R.Wallach, *On the unitarizability of derived functor modules*, Invent.Math. **78** (1984), 131-141.

[35] ———, *Real reductive groups I*, Academic Press, 1988.

[36] G.Zuckerman, *Tensor products of finite and infinite dimensional representations of semisimple Lie groups*, Ann. of Math. **106** (1977), 295-308.

TOSHIYUKI KOBAYASHI
DEPARTMENT OF MATHEMATICS
FACULTY OF SCIENCE
UNIVERSITY OF TOKYO
HONGO, TOKYO 113
JAPAN.

From April 1, 1991,

DEPARTMENT OF MATHEMATICS
COLLEGE OF ARTS AND SCIENCES
UNIVERSITY OF TOKYO
KOMABA, MEGURO, TOKYO 153
JAPAN.

MEMOIRS of the American Mathematical Society

SUBMISSION. This journal is designed particularly for long research papers (and groups of cognate papers) in pure and applied mathematics. The papers, in general, are longer than those in the TRANSACTIONS of the American Mathematical Society, with which it shares an editorial committee. Mathematical papers intended for publication in the Memoirs should be addressed to one of the editors:

General instructions to authors for

PREPARING REPRODUCTION COPY FOR MEMOIRS

> **For more detailed instructions send for AMS booklet, "A Guide for Authors of Memoirs."**
> **Write to Editorial Offices, American Mathematical Society, P.O. Box 6248,**
> **Providence, R.I. 02940-6248.**

MEMOIRS are printed by photo-offset from camera copy fully prepared by the author. This means that the finished book will look exactly like the copy submitted. Thus the author will want to use a good quality typewriter with a new, medium-inked black ribbon, and submit clean copy on the appropriate model paper.

Model Paper, provided at no cost by the AMS, is paper marked with blue lines that confine the copy to the appropriate size.

Special Characters may be filled in carefully freehand, using dense black ink, or **INSTANT** ("rub-on") **LETTERING** may be used. These may be available at a local art supply store.

Diagrams may be drawn in black ink either directly on the model sheet, or on a separate sheet and pasted with rubber cement into spaces left for them in the text. Ballpoint pen is not acceptable.

Page Headings (Running Heads) should be centered, in CAPITAL LETTERS (preferably), at the top of the page — just above the blue line and touching it.

LEFT-hand, EVEN-numbered pages should be headed with the AUTHOR'S NAME;

RIGHT-hand, ODD-numbered pages should be headed with the TITLE of the paper (in shortened form if necessary).

Exceptions: PAGE 1 and any other page that carries a display title require NO RUNNING HEADS.

Page Numbers should be at the top of the page, on the same line with the running heads.

LEFT-hand, EVEN numbers — flush with left margin;

RIGHT-hand, ODD numbers — flush with right margin.

Exceptions: PAGE 1 and any other page that carries a display title should have page number, centered below the text, on blue line provided.

FRONT MATTER PAGES should be numbered with Roman numerals (lower case), positioned below text in same manner as described above.

MEMOIRS FORMAT

> **It is suggested that the material be arranged in pages as indicated below.**
> **Note: Starred items (*) are requirements of publication.**

Front Matter (first pages in book, preceding main body of text).

Page i — *Title, *Author's name.

Page iii — Table of contents.

Page iv — *Abstract (at least 1 sentence and at most 300 words).

Key words and phrases, if desired. (A list which covers the content of the paper adequately enough to be useful for an information retrieval system.)

**1991 Mathematics Subject Classification*. This classification represents the primary and secondary subjects of the paper, and the scheme can be found in Annual Subject Indexes of MATHEMATICAL REVIEWS beginnning in 1990.

Page 1 — Preface, introduction, or any other matter not belonging in body of text.

Footnotes: *Received by the editor date.
Support information — grants, credits, etc.

First Page Following Introduction – Chapter Title (dropped 1 inch from top line, and centered). Beginning of Text.

Last Page (at bottom) – Author's affiliation.